St. Louis in the Century of
Henry Shaw

St. Louis in the Century of Henry Shaw

A View beyond the Garden Wall

Edited with an Introduction by Eric Sandweiss

Missouri Botanical Garden, Saint Louis
Missouri Historical Society, Saint Louis
University of Missouri Press, Columbia and London

Copyright © 2003 by
The Curators of the University of Missouri
University of Missouri Press, Columbia, Missouri 65201
Printed and bound in the United States of America
All rights reserved
5 4 3 2 1   07   06   05   04   03

Library of Congress Cataloging-in-Publication Data

St. Louis in the century of Henry Shaw : a view beyond the garden wall / edited with an introduction by Eric Sandweiss.
   p.   cm.
  ISBN 0-8262-1439-8
  1. Shaw, Henry, 1800–1889. 2. Shaw, Henry, 1800–1889—Influence. 3. Saint Louis (Mo.)—Biography.
4. Businessmen—Missouri—Saint Louis—Biography.
5. Philanthropists—Missouri—Saint Louis—Biography.
6. Saint Louis (Mo.)—Civilization—19th century.   7. Saint Louis (Mo.)—Social conditions—19th century.
I. Title: Saint Louis in the century of Henry Shaw.
II. Sandweiss, Eric.
F474.S253S53 2003
977.8'6603'092—dc21   2002015359

∞™ This paper meets the requirements of the
American National Standard for Permanence of Paper
for Printed Library Materials, Z39.48, 1984.

Designer: Jennifer Cropp
Typesetter: Bookcomp, Inc.
Printer and binder: Thomson-Shore, Inc.
Typefaces: Minion, Usherwood, Palace Script

Portions of the essay "African Americans in Henry Shaw's St. Louis" are reprinted from *Missouri's Black Heritage* by Lorenzo J. Greene, Gary R. Kremer, and Antonio F. Holland, revised and updated by Gary R. Kremer and Antonio F. Holland, by permission of the University of Missouri Press. © 1993 by the Curators of the University of Missouri.

For
Joy and Jerry Sandweiss

Contents

List of Illustrations   ix
Foreword by Peter H. Raven   xi
Foreword by Robert R. Archibald   xv
Acknowledgments   xvii

Introduction.
From a Garden Looking Out: Public Culture
in Henry Shaw's St. Louis   *1*
Eric Sandweiss

Part 1. Getting Along: Politics, Race, Ethnicity

Gods in Ruins: St. Louis Politicians and American
Destiny, 1764–1875   19
Kenneth H. Winn

African Americans in Henry Shaw's St. Louis   51
Antonio F. Holland

Learning from the "Majority-Minority" City: Immigration
in Nineteenth-Century St. Louis   79
Walter D. Kamphoefner

Part 2. Getting Ahead: Business, Science, Learning

The Economy of Nineteenth-Century St. Louis   103
James Neal Primm

Enterprise and Exchange: The Growth of Research
in Henry Shaw's St. Louis   136
Michael Long

Public Education in Nineteenth-Century St. Louis   167
William J. Reese

Part 3. Enriching Life: Theater and Literature

Shaping the Authentic: St. Louis Theater Culture and
the Construction of American Social Types, 1815–1860   191
Louis Gerteis

Themes and Schemes: The Literary Life
of Nineteenth-Century St. Louis   218
Lee Ann Sandweiss

Contributors   239
Index   241

Illustrations

Henry Shaw's Tower Grove House, late 1870s   opposite page 1
View of St. Louis, 1817   2
"St. Louis, from the Mississippi River," 1888   4
Missouri Botanical Garden, 1874   6
Henry Shaw's mausoleum, c. 1890   15
Thomas Hart Benton, c. 1856   23
Edward W. Bates, c. 1855   27
William Greenleaf Eliot, c. 1850   30
Frank P. Blair Jr., 1862   38
Inventory of Jeanette Forchet, May 30, 1790   54
Dred Scott, 1888 portrait after an undated photograph   61
Elizabeth Keckley, 1868   64
"The Cascades" sheet music cover, 1904   76
Sts. Peter and Paul Church, c. 1890   81
Witter's German-English Primer and New First German Reader for Public Schools, 1881   96
South St. Louis Turnverein, Young Ladies' Class Bar Drill, c. 1910   98
"Bird's-Eye View of St. Louis, Mo.," 1858   105
Bank of the State of Missouri $20 banknote, 1838   110
Advertisement for the Pacific Railroad, 1859   118
St. Louis Fairgrounds, 1859   122
William Clark, c. 1810   137
Dr. George Engelmann, c. 1855   142
Charles P. Chouteau, n.d.   154
Henry Shaw, c. 1880s   158
Shaw's Museum, 1868–1874   160
Missouri Botanical Garden entryway, c. 1875   162

William Torrey Harris, n.d.   171
Susan Blow, n.d.   176
Shepard Kindergarten, c. 1885   178
Manual Training Room, Hodgen School, 1899   185
Playbill for Dan Marble in *Jonathan in England* and *The Forest Rose*, 1849   196
James Henry Hackett as Nimrod Wildfire, n.d.   198
"Jim Crow, as Sung by Mr. T. D. Rice" sheet music, n.d.   206
Long Jakes; the Rocky Mountain Man, Charles Deas, 1844   213
Jolly Flatboatmen in Port, George Caleb Bingham, 1857   214
"Founding of St. Louis, 1764," 1902   220
William Wells Brown, 1880   226
"Terrible Tragedy at St. Louis, Mo.," 1861   231
Planters' House Hotel, c. 1870   233
Logan Uriah Reavis, c. 1875   235

Foreword

Peter H. Raven
Director, Missouri Botanical Garden

This book is about the city that Henry Shaw knew and helped to shape. It is the result of a collaboration between several of the institutions that Shaw founded or benefited during his lifetime: the Missouri Botanical Garden, Tower Grove Park, the Missouri Historical Society, and Washington University. The book's publication is also a collaboration—among the Garden, the Historical Society, and the University of Missouri Press. The essays in this volume were originally commissioned as contributions to a lecture-discussion series, held at the Garden in September and October 2000, which honored the two-hundredth anniversary of Shaw's birth on July 24, 1800.

Henry Shaw arrived in the United States as an immigrant from England early in 1819, when he was just eighteen years old. Acting as an agent for his uncle's ironmongering firm in Sheffield, Shaw used the knowledge of French he had gained at Mill Hill School to establish a depository in New Orleans, from which he shipped goods upriver. That spring, he arrived in St. Louis on one of the first steamboats to travel up the Mississippi River; in the coming years he supplied both the local population and thousands of westward-moving travelers with goods not otherwise available locally. When he died in 1889, Shaw left a legacy for all of us; it is that legacy that we celebrate in this book.

Shaw was a private man, but a man of action, and his life both reflected and influenced the exciting times that the talks in our series portrayed. Marshall Crosby, a botanist at the Missouri Botanical Garden, inaugurated the series with a portrayal of Shaw as seen through his letters and other writings in the Garden's extensive archives. As Crosby made clear, Shaw's successes exerted a significant influence on the city around him. By the time he ceased selling ironworks in 1840, he had amassed the then-considerable fortune of a quarter of a million dollars. He became in essence a private banker, acquiring and speculating in lands and other properties. In 1842, he gained possession of the lands that would ultimately become the sites of the Missouri Botanical Garden and Tower Grove Park.

It was at this point in his life that Shaw began to travel to England and other parts of Europe, especially around the Mediterranean, and it was then that he must have remembered his schoolboy love of plants and gardening. In 1851, during the summer of the Great Exhibition in London, he visited the estate of the Duke of Devonshire, at Chatsworth, Yorkshire, and was supposedly inspired by the greenhouse there (constructed by the architect Joseph Paxton, who also designed the Crystal Palace, the building that housed the Exhibition) to form a vision of a botanical garden in St. Louis.

When he returned to St. Louis in 1852, Shaw set about planning the kind of garden that he would build. He contacted the director of the Royal Botanic Gardens at Kew, Sir William Jackson Hooker, for advice; Hooker in turn referred him to Harvard professor Asa Gray, the most noted American botanist of the nineteenth century, and to Dr. George Engelmann, a German obstetrician who was an avid amateur botanist and a significant figure in the scientific and cultural life of St. Louis from his arrival in 1833 to his death in 1884. Engelmann became especially important in advising Shaw about his garden, but Hooker and Gray corresponded regularly and were key influences as well.

Shaw occupied his country home, Tower Grove House, in the autumn of 1850, and considered that the lands near it, some three miles outside the city at the time, would be a suitable place in which to develop his public garden. Aided by Engelmann, Gray, and Hooker, he did so, opening the Missouri Botanical Garden to the public in 1859. In 1867, he gave approximately three hundred acres of adjacent land to the city in return for bond financing that enabled him and his fellow commissioners to develop Tower Grove Park, and he immersed himself in planning the gazebos and plantings.

Shaw envisioned his garden as a display of the plants that would be useful in St. Louis and Missouri generally, a pleasant place to visit that would also serve as a place where people might learn about plants that they could cultivate themselves. Hooker, Gray, and Engelmann, however, convinced him that research—contributing to the store of knowledge about plants—was also fundamental to a complete botanical institution, and Shaw laid the foundation for the ample development of research that really began with the accession of William Trelease, hired four years earlier as professor of botany at Washington University, to the directorship of the Garden on Shaw's death in 1889. We do not know what sorts of educational activities took place at the Garden during Shaw's lifetime, but his appointment of the president of the St. Louis Board of Education to the Garden's board of trustees clearly indicated his serious interest in the Garden as an educational institution.

In addition to the Missouri Botanical Garden and Tower Grove Park, Shaw lent his energy and resources to many other institutions. He began the study of botany at Washington University with his establishment and endowment of a school of botany in 1885, when he was eighty-five years old. The linkage between Washington University and the Missouri Botanical Garden, which has proved so beneficial for both research and education, began in that year. Shaw was also involved with and helped establish St. Luke's Hospital (then located near his city home at Seventh and Locust Streets), the Missouri Historical Society, Christ Church Cathedral, the Mercantile Library, the Missouri School for the Blind, and a number of other local institutions. He believed in giving a measure of his wealth back to the community where he had earned it, and in so doing demonstrated well what Tocqueville described as a unique aspect of America: the voluntary association of members of society, putting aside their personal interests, to accomplish common goals for the betterment of the broader community. It is to that spirit—and to its broad impact on everyday life in nineteenth-century St. Louis—that this volume pays tribute.

Henry Shaw's accomplishments reflected the times in which he lived, which are analyzed in this collection of essays; but he also influenced those times and our own in ways that he neither planned nor could have foreseen. He was a man of the eighteenth century living in the nineteenth century, but his actions clearly continue to influence the lives of St. Louisans, and people throughout the world, in the early years of the twenty-first century, and will do so for the indefinite future.[1]

Although many others helped to plan and implement the activities that led to this volume, I am especially grateful to Robert Archibald and Eric Sandweiss and to John Karel of Tower Grove Park for their important contributions to the celebrations of the two-hundredth anniversary of Shaw's birth and to the planning of the symposium and subsequent development of this book.

---

1. By way of example, his leadership in establishing the Missouri Botanical Garden and what was later named the Henry Shaw School of Botany (now part of the Department of Biology) at Washington University was the catalytic force that has led, in our own time, to the formation of the Donald Danforth Plant Sciences Center and the effort among a variety of St. Louis institutions to develop a life-sciences-based economy for the twenty-first century.

Foreword

Robert R. Archibald
President, Missouri Historical Society

The garden in my backyard in south St. Louis is mostly bare on the winter day that I write this. The ivy still tumbles over the terrace; it's similar to the hardy variety my grandmother grew in Michigan. The rosebushes have been thoroughly pruned and mulched, and I know the roses will return this summer in all their lush and gory beauty. In the spring I'll plant tomatoes; my great-grandfather had rows and rows of vegetables, but I especially remember his tomato plants. My father liked small bright flowers, and some of the same are my regular choices, too. My garden, in any season, is more than just a pleasant haven and an enjoyable labor; it is also a mnemonic for recalling my forebears and the stories I know of them, the unanswered questions I have about their lives, the legacies and also the burdens that they left me.

While I would hardly compare this little patch of Pernod Avenue with the magnificent bounty of Henry Shaw's Missouri Botanical Garden, I can address some similarities. My garden, like Shaw's, is a legacy of several generations as well as a gift and a message to the community—in my case, to the neighbors and the strangers who pass by my home; in the Botanical Garden's, not only to its visitors and the people of the St. Louis region but also, in its work to improve and even save this planet, to the entire global community. As human beings we require a sense of something beyond ourselves and our own lives, a transcendence that admits to beauty and dignity and expectation. A garden, whatever its size and scope and scale, represents hope and a commitment to the future.

This is a commitment that we at the Missouri Historical Society share with the Missouri Botanical Garden and, I believe, with Henry Shaw himself. In his many charitable gifts and foundations, Shaw seemed to be looking toward the future but always with an awareness of the past, his own and his community's, and of the burdens and the benefits that previous generations had passed on. As I noted in my participation in the symposium for the two-hundredth anniversary of Shaw's birth that was the impetus for this book, the great majority of the institutional beneficiaries of his estate—seventeen of the nineteen listed in his will, including the Missouri Historical Society—are existing and flourishing more than a century later. "Shaw's Garden" has achieved worldwide honor and contributed immensely to the improvement of the community and the entire planet. These lasting achievements are the result not of a dream but of a plan that succeeded abundantly. That he was a man of vision is so apparent that the phrase is nearly disparaging. Yet a "man of vision" he was.

As a teenaged immigrant, Shaw decided to join his future with the frontier town of St. Louis. Very much a man of his times and his chosen place, he mapped out his road to success and, his own success achieved, concerned himself with the community that had provided the opportunity for his prosperity. Among his legacies to St. Louis and to the world, surely that example can serve us well in an era where "place" and "community" have become extricable from one another.

Shaw's life nearly spanned the nineteenth century, and his influence has spanned the twentieth and continues into the twenty-first. Thus the stories of his era and his chosen place are stories that connect us to the past and ultimately connect us with each other.

Henry Shaw was, as I have mentioned, a man of his times, and those times, seemingly long ago and far away and very different from our own, bear examination as more than an antiquated curiosity. The people who lived here in the nineteenth century had much in common with us. They too felt a need for beauty and dignity and transcendence. They worried about providing for themselves and cared about how they got along with one another. They knew love and laughter, sorrow and celebration, the pleasure of a spring garden and the chill of a winter night. They planted, in their own ways, the seeds of the future, a future we are now living as we ourselves become part of the past. In examining and evaluating their times, we can discern how past, present, and future interact and learn how choices we make, seeds we plant, affect those coming after us.

In a most exciting and appropriate collaboration, we are pleased to look "beyond the garden wall" to St. Louis and the multiple aspects of its life in the 1800s.

Acknowledgments

In a manner that Henry Shaw would no doubt have approved, this book grew out of the cooperative efforts of a number of individuals and organizations. First and foremost, it owes its existence to the Missouri Botanical Garden, which chose in 2000 to commemorate the two-hundredth anniversary of the birth of its founder, Henry Shaw, with a series of public programs and commemorations. Director Peter Raven, working with Deputy Director Jonathan Kleinbard, oversaw this yearlong birthday celebration with an admirable combination of broad vision and close attention to detail. They also brought together a powerful team of individuals to flesh out the details of the symposium that focused on St. Louis in Shaw's time. This team included John Karel, the director of Tower Grove Park and a distinguished historian in his own right; Lynn Hagee, who undertook the considerable task of locating and contacting the speakers whose talks were eventually transformed into these essays; and Marshall Crosby, who lent his expertise as the Garden's chief in-house expert on the history of the Garden and of Shaw's life. Archivist Andrew Colligan later provided invaluable assistance with images.

The Missouri Historical Society (MHS), one of the beneficiaries of Shaw's largesse in the nineteenth century, proved a fitting partner in the project, contributing staff time and resources from the stage of initial program planning through the realization of the book. The influence of MHS President Bob Archibald turned the dream into reality, while Director of Publications Lee Ann Sandweiss performed triple duty, serving at once as the project manager for the book, as one of its contributors, and—most laborious of all—as the wife of its editor. Lauren Mitchell saw the project through on behalf of the MHS Press, Duane Sneddeker and Ellen Thomasson shared their tremendous knowledge of MHS's print and photograph collection, and Katie Hunn provided key assistance in image research and selection.

The final partner involved in the production of this book is the University of Missouri Press, whose director, Beverly Jarrett, graciously approved the manuscript for publication and helped to sharpen its content along the way. Editors Jane Lago and Gary Kass are responsible for giving coherent form to a set of disparate essays—a difficult task for any editor, and one that, if successful, goes unnoticed by unsuspecting readers.

Finally, I thank the distinguished contributors to this volume for lending their expertise, their erudition, and especially their patience to the project. It has been an honor to learn from them, and a pleasure to bring their knowledge together into one volume from which I hope historians and interested St. Louisans will benefit for years to come.

St. Louis in the Century of
Henry Shaw

Henry Shaw's Tower Grove House, late 1870s. Shaw spent the final three decades of his life at this country estate, located on land that he dedicated for use as the Missouri Botanical Garden. With Shaw in the image is Rebecca Edom, his housekeeper. The view is to the southeast and shows the original east wing, which was replaced in 1891. No other exterior additions have been added since. *Missouri Historical Society*

Introduction

From a Garden Looking Out

Public Culture in Henry Shaw's St. Louis

Eric Sandweiss

In 1869, on the occasion of the fiftieth anniversary of his arrival in St. Louis, Henry Shaw invited 175 friends and fellow citizens to the Missouri Botanical Garden to hear his reflections on a long and profitable life in the gateway to the West. Shaw was, by this time, known not only to his well-heeled dinner companions but to virtually all of the residents of this fast-growing city. A British immigrant who built one of St. Louis's major private fortunes through his hardware trade, he had long since turned from his business pursuits to lead the life of a self-styled philanthropist. From a deep sense of civic duty, as well as from a wide-ranging interest in human affairs, Shaw had already, by this date, begun to leave a conspicuous legacy of gifts to his adopted city, including the two that remain most recognizable in our own time: the Garden itself and, bordering it on the south, the newly dedicated Tower Grove Park.

Shaw had little to say that evening, however, about his business successes or about his bequests to the public landscape of St. Louis. Instead, he chose to look back on the day, a half-century earlier, when he first sighted the town from his perch on the deck of a northbound steamboat. Through the haze of the passing years, Shaw recalled the town's "cheerful appearance" on that May morning, "some of the houses being elegantly built with wide verandas, in the Louisiana style." While he saw along the waterfront little of the commerce on which the city would subsequently depend, the amateur botanist remembered that "on the top of the bank were gardens, with fruit trees in blossom, forming a pleasant prospect, compared to the swampy lands and moss-covered trees of the Lower Mississippi." Tied, in his mind, to this pleasing and gracious landscape were the families who dwelt within it—families with names like Cabanne, Gratiot, and Soulard—whose "kindness, courtesy, and politeness" Shaw implicitly opposed to the manners common to the Americanized, rough-edged city of a half-century later.[1]

---

1. "The Enigma of Mr. Shaw: In Commemoration of the Centennial of the Missouri Botanical Garden," *Bulletin of the Missouri Historical Society* 15:4 (July 1959): 312.

View of St. Louis, 1817 (woodcut). This earliest known view of St. Louis, imprinted on a locally issued banknote, captures the town much as it appeared when Henry Shaw arrived from New Orleans in 1819. Fifty years later, Shaw still recalled the spring day when he first glimpsed St. Louis's waterfront from the deck of a steamboat. *Collection of Eric P. Newman*

Shaw was not alone among St. Louisans in waxing nostalgic for simpler and quieter days. Such sentiments were common enough in the 1860s—by which time the city had already weathered epidemic, flood, and fire, decades of immigration, and five years of civil war. Indeed, it was just three years before Shaw's talk that he and a few of his acquaintances founded the Missouri Historical Society, hoping thereby to preserve for future generations some trace of the nearly vanished world that Shaw and others still recalled well. The posthumous discovery and publication, several years prior to that, of Auguste Chouteau's narrative of the founding of the city in 1764 had likewise helped to stimulate a broadening sense of civic memory and, with it, a longing among many St. Louisans for the simplicity of a smaller and far less heterogeneous community.

What surprises us today in reading Shaw's remarks is not his fond recollection of St. Louis's pre-urban roots, nor the considerable degree to which those roots had been obscured in the fifty years since his arrival. More striking, in hindsight, is Shaw's (and his audience's) necessary ignorance of how much the city would *continue* to change in the two decades that remained of his already long life. By the time he died in bed at Tower Grove, his country estate, in 1889, St. Louisans were already talking about hosting a World's Fair. Everywhere, the city was growing: the fields around Tower Grove, far from downtown, were marked with construction sites and surveyors' stakes, and the Grand Avenue viaduct, a mile east, would soon bring a steady stream of streetcars and wagons from densely packed neighborhoods in the city center toward south St. Louis.

Far from bearing passive witness to such changes, Shaw had played an active role in bringing them to pass.

Of Shaw's life it can truly be said, as the *St. Louis Republic* noted after his passing, that "his career in this city is coincident with the city's development." Shaw arrived in the city shortly after the first steamboat; he died as electric lines were being installed above the streetcar tracks and the tower of Union Station was beginning to rise above Market Street. The duration of a trip from New Orleans, such as the one Shaw himself made, had gone from nearly a month to just over a day. The town whose resources could be precisely inventoried in a single essay, such as John Paxton prepared in 1821 (651 houses, 57 grocers, 2 brickyards, 5 billiard tables, etc.), was now the topic of a growing number of massive books, all of which, taken together, could never hope to succeed in matching Paxton's comprehensiveness.[2]

Indeed, Shaw's long life, more than most any other that one can conjure, cut across an extraordinary range of St. Louis history and St. Louis characters: it overlapped the lifetimes of both Daniel Boone and T. S. Eliot, of Auguste Chouteau and Branch Rickey. Just before the moment of his death, on an August night in 1889, Shaw is said to have uttered the word "seventy-nine"—an apparent reference to what he believed to be his advanced age. It is little wonder, at a time when the average life expectancy of an American man was far less than it is today, that Shaw undershot the mark by a full decade; to imagine that his life had truly lasted for nearly a century would have strained the credulity even of those faithful few who attended him at the end.

But speaking of his own life—even in such clipped terms—was not Shaw's style. Given the opportunity, on that May evening in 1869, to summarize his achievements, Shaw chose, characteristically, to focus his address instead on the changes seen in the city that lay beyond the Garden's high limestone walls. Much as we know of nineteenth-century St. Louis from his and countless others' recollections, Shaw himself was, and remains today, something of a cipher. His biographer, William Barnaby Faherty, who has given us the most complete portrait now available of this driven man, draws a picture of an individual whose sense of public obligation served in part to mask a passion for privacy. Never lacking for friends, he nevertheless seems to have compartmentalized his personal relationships, building them around common interests—gardening, wine, business—rather than from a desire for general companionship. A bachelor to his death, Shaw left only occasional references in his letters to the many

---

2. *St. Louis Republic,* August 26, 1889; John A. Paxton, "Notes on St. Louis," in *The Early Histories of St. Louis,* ed. John Francis McDermott (St. Louis: St. Louis Historical Documents Foundation, 1952), 67–68.

"St. Louis, from the Mississippi River. View of the levee from the Bridge" (wood engraving after Charles Graham from *Harper's Weekly,* June 9, 1888). By the time of Shaw's death in 1889, St. Louis, approaching a half million in population, was one of the largest cities in North America. The waterfront, no longer the pastoral mixture of residences and public buildings that Shaw and others had remembered, now reflected the industrialization and commercial expansion that fueled the city's tremendous growth. *Missouri Historical Society*

women who must have sought his company; his most noteworthy romantic episode was, in fact, the sensational 1859 breach of promise suit brought by Effie Carstang, whose $100,000 award (subsequently overturned on appeal) brought an unwelcome sort of national press attention to the retiring entrepreneur. The most constant female presence in Shaw's life was his sister Caroline—like him, unmarried—who helped him manage his numerous investments over the years.[3]

Further evidence, indirect though it may be, also suggests that Shaw's business drive and civic commitment helped him to divert his and others' attention away from his private affairs. Henry and Caroline, with another sister, Sarah, were shadowed through much of their lives by the business failures of their father, a Sheffield merchant with a knack for ill-considered speculation. It was in order to manage one such venture (as well as, perhaps, to escape his father's grip) that Shaw came to the United States in the first place, following the family's hardware shipment to New Orleans, where it was languishing without a buyer. As Shaw succeeded in putting more miles between himself and his father (passing up numerous opportunities either to return to him or to bring him to St. Louis), he also succeeded in transcending his father's bad luck in business.

3. William Barnaby Faherty, *Henry Shaw, His Life and Legacies* (Columbia: University of Missouri Press, 1987), 30–31.

The risky decision to bring that first shipment of goods from the busy port of New Orleans to the distant backwater of St. Louis—a decision based more on St. Louis's promise as a future entrepôt than on its proven record as a center of commerce—reflected a measured aggressiveness that would serve Shaw well as he went on to amass one of the city's largest fortunes. Surviving the financial panics of 1821 (when he had little to lose) and 1837 (when he had much), he funneled ever-greater amounts of capital away from the commodities that had given him his financial start and into longer-term investments. By 1839, according to Faherty, fully half of Shaw's already substantial estate consisted of mortgages and bonds.[4] In the years that followed, he divested himself still further of the hardware business, buying up more properties from which to draw a steady rental income.

As Shaw aged, he used that growing income to finance an increasingly expansive—though never extravagant—lifestyle. He returned to Europe in 1840–1841, then again a decade later. As Peter Raven notes in his foreword to this volume, Shaw did more than savor the usual delights of the Grand Tour. He also focused his attention on the continent's best-known examples of botany and landscape design, familiarizing himself with the Royal Botanic Gardens at Kew and making the acquaintance of such leading figures in the field as Joseph Paxton and William Hooker. From these visits he acquired the knowledge that would serve him so well in the pursuits that occupied the second half of his life: principally, the development of the Garden and of Tower Grove Park, which opened in 1859 and 1868, respectively. Between the two European sojourns, Shaw engaged the services of St. Louis's best-known architect, George I. Barnett, to build both his massive townhouse downtown and his country estate, Tower Grove, on the grounds of what soon became the Garden. Through these well-known structures, as through his great public land bequests, he established himself as a presence to be reckoned with in the American West's largest city.

If Shaw increasingly became a civic individual, cutting an ever-wider figure in nineteenth-century St. Louis, he nevertheless demonstrated the same reserve and elusiveness in public affairs as he did in his private life. While Faherty describes an early and apparently sincere repugnance for what Shaw termed, in one letter, the "knavery, oppression, and slavery" of 1820s Missouri, Shaw's correspondence and personal behavior demonstrated his willingness to look the other way when it suited his purpose. "Politics claim a very small part of my attention," he asserted in another letter, not many years later—and, indeed, he was sufficiently able to distance his own actions from their ethical and political

---

4. Ibid., 39.

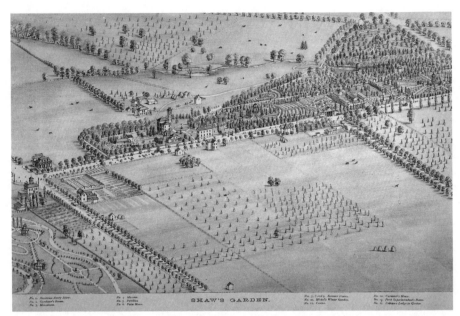

Missouri Botanical Garden (lithograph by Camille N. Dry, in Richard J. Compton and Camille N. Dry, *Pictorial St. Louis* [St. Louis: Compton and Co., 1874], plate 91). By 1874, a few years after Shaw invited prominent St. Louisans to celebrate his fiftieth anniversary in the city, the Garden was already taking the shape familiar to St. Louisans today. This bird's-eye view—from a volume documenting every block in the city of St. Louis—shows the landscaping beginning to develop around Shaw's country home. *Missouri Historical Society*

repercussions that he spent three decades as a slaveowner, even hiring men to chase and return those slaves who dared escape. Shaw seemed to consider the slave issue strictly a business matter, and his long record as an owner of human flesh did not deter him from adopting the pro-Union stance of most of his friends and business colleagues at the time of the Civil War. Like most successful capitalists, Shaw was also loathe to consider that there might be a political dimension to the economic system on which he depended for his success. Instead, in his public writing, he dismissed organized labor's attempts at improving the lot of the workingman as a stultifying force, "annihilating all incentive to individual excellence and perfection."[5]

It was Shaw's own relentless drive for "individual excellence and perfection" that seems to have accounted, as much as any particular passion for learning, for his success as a promoter of botanical knowledge. The physician and botanist

---

5. Ibid., 14, 30; Henry Shaw, *The Vine and Civilization* (St. Louis: Tower Grove, 1884), 17.

George Engelmann, who benefited immensely from Shaw's generosity and who as a longtime associate had a greater opportunity than virtually anyone to understand the inner workings of his mind, never doubted Shaw's sincerity or effectiveness. Still he lamented of his patron: "would that he had more scientific education or taste."[6]

Politics, science, education, taste—these are among the areas in which Shaw distinguished himself less by his personal contributions than by an instinct for recognizing and promoting the important innovators of his time. To find such innovators, he could scarcely have come to a better place than St. Louis, a city that in the 1800s drew to it a remarkable cast of players and provided them with a conspicuous stage on which to work. This is a book about that city, a place whose development, as the editors of the *Republic* wrote on the occasion of Shaw's passing, coincided with (and benefited from) his remarkable life. In the chapters that follow, Henry Shaw will appear, vanish, and reappear, making felt his ubiquitous presence in every aspect of the city's culture, just as he did in life. But ultimately, in a manner that this civic-minded man would likely have approved, Shaw is less the focus of these essays than is the world that he knew—the city that grew, between 1800 and 1900, from that remembered village with its "cheerful appearance" to the largest metropolis in the American West, a place of skyscrapers and steel factories, ragtime dance halls and German beer gardens, kosher groceries and Gilded Age mansions.

While Shaw's life has been admirably accounted for by Faherty and other historians, the far more diffuse topic of the culture of the remarkable city that he helped to shape still wants for a comprehensive survey. This book seeks to make real for twenty-first-century readers the link between man and city that was second nature to most nineteenth-century St. Louisans. Shaw's life provides us with a lens through which to begin to envision what might otherwise seem the almost impossible task of accounting for the city's vibrancy. Though its history is less widely known than that of the nation's other major postbellum cities (Philadelphia, New York, Boston, Baltimore, and Chicago), St. Louis—and the people and events that came together there in the 1800s—contributed actively to the development of many of the features of American society that we live with today. Because Americans and foreigners alike recognized St. Louis's important place in the unfolding story of a hybrid and original American culture, the city's renown extended across the nation and around the globe. Shaw was only one among many who came there from far away and whose deeds would in turn go on to affect people across the nation and abroad.

---

6. Faherty, *Henry Shaw*, 16.

More than a city, then, St. Louis was a world in itself—as visitors never failed to remark. The city's long-noted status as a "gateway"—to distant points, to opportunities unknown at home—attracted to it an astonishing range of individuals with a surprising variety of intentions. Making their way up the slope of the levee in the years after Shaw first gazed upon it were American Indians from both sides of the Mississippi, free and enslaved blacks, migrants from all sections of the Eastern United States, and a notable concentration of foreign immigrants, ranging from the British and Irish, who were dominant early in the century, to the Italians, Chinese, Poles, and Russians who arrived later on. Throughout those years, of course, no ethnic group made as conspicuous a mark as the Germans, whose language and customs prevailed in many parts of the city for much of the century. But regardless of where one turned in St. Louis, the city's location at a key break in North America's inland water system—at a point that was culturally and geographically neither wholly South nor North, East nor West—gave to it the cosmopolitan character that travelers and newcomers often noted with surprise.

It is in seeking to explain and evoke that character that I have to rest upon a vague but unavoidable term, already introduced above: *culture*. We tend to use the word, in our everyday conversation, to refer to those resources and activities that are supposed to be "good for us": enriching experiences that take us away from our concerns with the necessities of life and put us in touch with a higher plane of artistic, ethical, or spiritual experience. But in the sense in which this book will explore the word, *culture* is a more encompassing term. It refers to the ways in which people mediate their experiences with one another—the shared traditions, words, and codes of behavior that allow individuals to retain a sense of their own identity while taking advantage of the safety, wealth, and stimulation that belonging to a larger community affords them. Creating and defining a shared culture is the essential task of a city, where, by definition, people must learn to live beside others whom they do not know. A man like Henry Shaw, who came into contact with great numbers of individuals from around the world, was, in this sense, a key cultural figure, far beyond his involvement in such "cultural" interests as parks and education. It is for this reason that his life and his works provide such a good entry point into the topic of the city as a whole.

Culture in the sense in which it is treated in this book, then, includes a number of areas that may seem to some readers to belong outside of the cultural arena. It includes, for instance, politics—which can be seen, after all, as sets of rules and expectations for expressing and negotiating the desires of many individuals seeking to benefit from shared resources. It includes the habits and institutions developed around affiliations of skin color, gender, national or regional origin, and religious belief—affiliations that both shape and are shaped by the perceptions and actions that groups of people visit upon one another. Culture is the interaction of the banker with his borrower, the attorney with the judge, and the editor with his readers, just as surely as it is the flair of the architect's pen or the resonance of the actor's voice. In each case, a web of shared memories, expectations, language, and behavior moderates and regulates what might otherwise be a meaningless personal interchange. In the details of these personal interchanges, in addition, lie the means by which such shared expectations can be contested or changed in the future.

It is in cities, and particularly in fast-growing, fast-changing cities like nineteenth-century St. Louis, that the relationship between shared conventions and personal creativity—and therefore the very nature of public culture—is at its most dynamic and unsettled. It is in cities, as Gunther Barth shows in his important study of urban culture, that people regularly develop "new patterns of getting along with one another."[7] Here, convention and innovation constantly butt up against one another—sometimes to creative and sometimes to destructive effect. Shaw was but one of thousands in St. Louis who brought with them the habits they had known from a distant land, adapting them to what they encountered on arrival at this new place. His identity and his personality, like those of his fellow citizens in nineteenth-century St. Louis, were already formed on his arrival from England—but only to a point. They would be provoked, redirected, and transformed in the decades he spent in this fluid city; inevitably, just as he helped to transform the culture of St. Louis, he left it, on his death, a very different man than he had been on the day of his arrival.

It is this unstable tension between individual and society—so different in its nature from the stable republican community envisioned by the founders of the United States—that captured the attention of American intellectuals and the concern of political and business leaders in the late 1800s. If unpredictable cities like Shaw's St. Louis—rather than the farms of Virginia, say, or the villages of New England—were to prove the forge in which an American character was molded, then, they reasoned, it was time to rethink seriously what kind of nation this was turning out to be. Economist Adna Weber, noting that "the most remarkable social phenomenon of the present century is the concentration of population in cities," reflected many Americans' uncertainty as to whether the unpredictability of city life was a phenomenon to be invited or avoided. On the one hand, wrote Weber, city people were capable of "a broader and freer judgment and a great inclination to and appreciation of new thoughts, manners, and

---

7. Gunther Barth, *City People: The Rise of Modern City Culture in Nineteenth-Century America* (New York: Oxford University Press, 1980), 4.

ideals"; on the other, that freedom listed dangerously toward "extreme individualism" and "class antagonism." Arthur Schlesinger Sr., writing a generation later, captured the puzzlement of nineteenth-century Americans concerning the city in a simple question: "Was its mission to be that of a new Jerusalem or of ancient Babylon?"[8]

As the essays in this book make clear, St. Louisans with a stake in their city's future worked diligently to see that that question would be answered by the first of the two alternatives. Whether possessed of a true religious zeal (like William Greenleaf Eliot, one of the principal characters in Kenneth Winn's essay on the city's political scene) or the faith of such avid capitalists as Thomas Allen (highlighted here by James Neal Primm), or simply driven by visions of global greatness (as Lee Ann Sandweiss demonstrates of civic booster Logan Reavis), powerful St. Louisans saw their mission as one of leading the city toward its lofty destiny. Equally clear from what we learn here, however, is the fact that these influential men—to say nothing of the vast numbers of citizens whose voices are less accessible to us today—often disagreed about what such greatness should consist of (the city's divided status during the Civil War, which figures in much of what follows, is only the most dramatic example of such disagreement). If nineteenth-century St. Louis was a place of extraordinary leaders—many of whom appear in these pages—it was also a place of extraordinary contention. It is this contentiousness that makes this city such a good place from which to view the conflicts through which all American cities, and the nation itself, would define themselves in the transition from a new republic to an industrial empire.

In order to give some structure to the elements of public culture in nineteenth-century St. Louis, I have chosen to follow the organization of the series of public programs upon which these essays are based. Part 1, "Getting Along," focuses on the challenges faced by people seeking to reconcile their differences in the diversifying city. Kenneth Winn's essay examines the careers and relationships of St. Louis's (and Missouri's) most illustrious political leaders—men like Thomas Hart Benton and Edward Bates—to develop an understanding of the fractious and often confusing context in which St. Louis distinguished itself from the state around it and in which Missouri itself came to occupy a key position in the national political debate that led up to, and then out of, the Civil War. Winn's essay points to the unique convergence of political passions,

---

8. Adna Ferrin Weber, *The Growth of Cities in the Nineteenth Century, A Study in Statistics* (New York, Macmillan, 1899), 432–34; Arthur Meier Schlesinger, *The Rise of the City, 1878–1898* (New York: Macmillan, 1933), 443.

personal ambitions, and even family relationships that made Shaw's St. Louis a place of such extraordinary political import.

Antonio Holland and Walter Kamphoefner take up the challenge of describing the place in this political maelstrom of two of the city's numerous ethnic communities: African Americans and German Americans, respectively. As their essays show, both groups played a crucial role in St. Louis's split with secession-leaning Missourians in the Civil War. More broadly, Holland and Kamphoefner examine how such self-identified communities either chose or were forced to retain distinctive cultural and political means for reinforcing their own solidarity in the city—even as they were engaged in "becoming St. Louisans." Far from being a simple matter of either assimilation or exclusion, the process of getting along as a minority—or even as a "majority-minority" community, in Kamphoefner's phrase—is revealed in these essays to be a process of negotiation, one that is heavily influenced by the particularities of St. Louis's "gateway" status in the middle of the continent. As German ethnics became less segregated and African Americans more so in the course of the century, these two groups illustrated the tremendous complexities of the process by which cultural distinctions are cultivated or overcome within a diverse urban environment.

Part 2, "Getting Ahead," focuses on those elements of St. Louis's culture that were forged around the task of advancing its people, either through economic or intellectual betterment. James Neal Primm sets the scene with his overview of the city's developing economy, in which he describes the sweeping transition from the city's early days as an outpost in an international fur-exchange network to its development as an industrial manufacturing center. Primm continues the lines of inquiry that he pioneered in his career as a historian, leading us through the complex intersections of banking policy, transportation technology, and national and international investment patterns in order to contextualize the lives both of successful entrepreneurs like Shaw and of the hundreds of thousands of St. Louisans whose labors earned them little more than food on the table at day's end.

In the essays that follow Primm's, we learn that St. Louis's civic and business leaders (most notably, Shaw himself) considered the task of educational or cultural advancement to be one with the city's economic advancement. Michael Long outlines St. Louis's role as a center for scientific exploration and study, weaving the story of the origins of Shaw's Missouri Botanical Garden into that of the Academy of Science, which developed because of the city's fame as a jumping-off point for western exploration. As Long shows, St. Louis's good fortune in bringing together investment and intellect placed it on a par with such longer-established cultural centers as Philadelphia in terms of its importance as a site for international scientific research. In his essay, William Reese emphasizes the city's equal prominence in the development of American public education. Focusing on the three extraordinary figures of William Torrey Harris, Calvin Woodward, and Susan Blow, Reese shows how the philosophy of public schooling—a vital element of the evolving definition of American citizenship—was articulated more sharply in St. Louis than anywhere else in the country. His research reveals intense personal and philosophical links between these innovators, even as it delineates the fundamental differences that sometimes kept them apart (and that continue to shadow the discussion of public education to this day).

In Part 3, "Enriching Life," we visit the area that most closely corresponds to our own traditional understanding of "culture." The essays of Louis Gerteis and Lee Ann Sandweiss make clear the importance of artistic and imaginative life to the everyday landscape of the city; further, they position St. Louis, once again, as a center for individuals and movements that would go on to have national influence. Gerteis looks at the city's active theatrical life in terms of the stage characters and stereotypes that helped St. Louis audiences come to terms with the burgeoning diversity that they saw around them each day. Into these theatrical types (and the artistic images they inspired)—sometimes cruel, sometimes generous, but always packed with a significance that transcended their function as mere entertainment—St. Louisans channeled their anxiety over their fast-changing city.

Likewise expressive of the growing pains of the time were St. Louis's contributions to literature, which Sandweiss explores across a wide range of written material, ranging from early-nineteenth-century Indian captivity narratives to the celebrated pages of William Marion Reedy's *Mirror* at the century's close. Sandweiss demonstrates that the city's "frontier" character did indeed shape its literature, but that an even stronger strain of self-conscious civic destiny eventually developed as the thematic backbone of much of the city's considerable literary output. In literature, as in theater, the city's reputation as a funnel—through which ideas and individuals from around the world were squeezed together into a shared civic culture—helped to garner for St. Louis a cultural variety and richness that belied its physical distance from older cities. The problem of defining and preserving that shared culture in the Gateway City became as much the charge of writers as it did of politicians and business leaders.

These, then, are the most important aspects of the world that Henry Shaw knew and helped to shape. Unquestionably, the city he left behind—a city pointed squarely in the direction of the twentieth century—was vastly different from the one he had come to on that May morning in 1819. Much has been written about St. Louis at the turn of the twentieth century—a time when the city ranked fourth in the nation in its population, when a World's Fair brought the arrival of twenty million visitors and the attention of countless more, and when the realization of Logan Reavis's hyperbolic prediction of St. Louis as the "Future Great City of the World" seemed to be just around the corner. For more than a decade, since shortly after Shaw's death, the determined phrase "New St. Louis" had served as the slogan of civic leaders pushing for change and investment.[9]

But how much of St. Louis was in fact "new" at the turn of the century? How much can a city's public culture serve to erase the sites, memories, and habits of the past? For all the insistence with which boosters clung to their program of urban reform and development, the atmosphere of an old St. Louis, a St. Louis shaped by the kinds of people who appear in this book, still hung palpably in the air. Three years after Shaw's death, the young reporter Theodore Dreiser found that atmosphere in his walks through the city's streets, where he saw "old buildings . . . all brick and all crowded together, with solid wooden or iron shutters and windows composed of such very small panes of glass. Their windows were so dark, so redolent of an old time life."[10] Reedy, who would support Dreiser's leap from reportage to fiction, reflected the feelings of those St. Louisans who saw little to cheer about in such scenes. For Reedy and others, the obverse of the "New St. Louis" for which they pushed was the reality of the St. Louis that they still saw around them. Reedy's "What's the Matter with St. Louis?" editorial, discussed by Sandweiss in her essay, reflected the fear among many that the spirit of progress and public welfare that once seemed so inexorable to the men of Henry Shaw's generation might in fact have died out with the philanthropist himself.

The "New St. Louis" hyperbole had always reflected a widely shared insecurity that the city's bucolic past and closely linked circles of social and political power, so fondly remembered by men like Shaw, might now be serving as an impediment, rather than a spur, to the city's ability to keep its edge in the new century. Yet the boosterish talk itself gave rise to an anxiety of a new

---

9. *The New St. Louis: Illustrated from Photos by Fred Graf* (St. Louis: Fred Graf, 1902). See also Ernest D. Kargau, *Mercantile, Industrial, and Professional St. Louis* (St. Louis: Nixon-Jones, 1902). For historical perspectives on the political appropriation of the "New St. Louis" phrase, see James Neal Primm, *Lion of the Valley: St. Louis, Missouri, 1764–1980*, 3d ed. (St. Louis: Missouri Historical Society Press, 1998), 363, 373; and Eric Sandweiss, *St. Louis: The Evolution of an American Urban Landscape* (Philadelphia: Temple University Press, 2001), 190.

10. Theodore Dreiser, *A Book about Myself* (New York: Boni & Liveright, 1922), 93.

and different sort: the anxiety that the city might in fact be losing its "culture," its connection to the hard-won, shared traditions—many of them profiled in this book—that had once made it unique. As street scenes like the one that Dreiser described became increasingly rare, as new immigrants changed the social makeup of the city yet again and new levels of corporate organization brought greater regimentation to people's lives, a generation that had never known the village witnessed firsthand by men such as Shaw would seek to re-create for its own purposes a sense of the past.

While it would remain for the great Pageant and Masque of 1914—a three-day spectacle of history and drama in Forest Park—to give to this urge its fullest expression, some St. Louisans sensed the change even by the time of Shaw's passing. Harry Turner, litterateur and socialite manqué, was among the first to complain of the cost of success. "We no longer like to be called a town," wrote Turner in the 1890s. "We are a city now, and have doffed our simple ways and are disdainful of . . . childish and fatuous pursuits. . . . We have become a serious people, with a serious object in life (though just what that object is, is a never-ending subject for discussion, or more properly, dissension). We talk now of 'Home Rule,' 'Social Evil,' 'Free Bridge,' 'A New Charter,' or 'A Million Population' . . . To not be in a state of hysterical excitement about one or the other of these projects is to argue that one lacks civic pride and is without a mission or purpose in life."[11] Such talk was, of course, simply the evidence of a grasping toward a new sort of public culture, a new set of rules and habits for conceiving of the city and the individuals who lived in it. While one may rush to conclude, with Turner, that St. Louis had "lost" the culture profiled in this book, a culture that once made it so distinctive in the American eye, the city was, like every other metropolitan center, simply adjusting itself to the changing world, adapting what had come before to what was yet to come.

Henry Shaw, despite his typically nineteenth-century commitment to development and "progress," never saw such change as necessitating the eradication of the past. In his treatise on the rose, one of several small books penned in the final years of his remarkable life, Shaw took time out from his historical discourse for an uncharacteristically personal reflection, one that speaks to the depths both of the man and of the city he loved. "Some leaves," he marveled, "gathered by the writer at Tower Grove, in 1852, and preserved in a jar, are now, (1877), still fragrant."[12] St. Louisans today, looking uncertainly at yet another new century, can take heart from the fact that the deeds and words of Henry Shaw—and of all the people who appear in this book—still carry, like

---

11. Harry Turner, *Casual Moments* (St. Louis: Greeley Printery, n.d.), 239.
12. Henry Shaw, *The Rose* (St. Louis, R. P. Studley & Co., 1879), 11.

Henry Shaw's mausoleum (photograph by Emil Boehl, c. 1890). Henry Shaw's funeral was a highly publicized event in the city to which he had dedicated his philanthropic energies for a half-century. Like the specially designed tomb in which he was interred on the grounds of the Garden, the evidence of Shaw's impact upon his adopted city remains visible today. Standing beside the tomb is the wife of William Trelease, the man who succeeded Shaw as director of the Garden. *Missouri Historical Society*

the essence of that Tower Grove rose, across the years. As Shaw himself recognized in his remarks on the city's changes across fifty years of his life, the lingering presence of our past burdens us in ways we cannot avoid. But it also blesses us—blesses us with great reserves of experience from which we can draw in shaping a culture that carries us into the future.

Part 1

Getting Along

Politics, Race, Ethnicity

Gods in Ruins

St. Louis Politicians and American Destiny, 1764–1875

Kenneth H. Winn

In the beginning all of the world was America. —John Locke

A man is a god in ruins. —Ralph Waldo Emerson

In the beginning all of the West was St. Louis. From its founding in 1764, people struggled over what the village, the town, and, eventually, the city would become. St. Louis was famously shaped by its Mississippi-Missouri river geography, but its politicians, no less than its powerful rivers, attempted to mold the city according to their own desires. Many of these politicians believed, during the years in which Henry Shaw was finding the city a perfect site in which to earn his fortune, that the West would ultimately determine America's character and destiny, and that St. Louis would control the American West. While many of their fellow citizens proclaimed the inevitability of St. Louis's future as the nation's greatest city, most politicians, leaving nothing to chance, worked feverishly for the city's coming glory. With the character of St. Louis, the state of Missouri, and ultimately the nation up for grabs, Missouri politicians contested for power with great seriousness.

The political stakes involved in shaping the city grew over time, and as they did, the difficulty of exercising control only increased. At the time of its founding, St. Louis served principally as a launchpad for hunting and Indian trading expeditions to the West. Meaningful political control resided in New Orleans. As the founding French elite reluctantly gave way to Americans after the Louisiana Purchase, the growing town's leadership fell to a group of young, aggressive settlers, a number of them western expansionists who envisioned an agrarian empire for which St. Louis would serve as headquarters. Beginning around 1840, however, the city's population exploded and everything changed. Fueled by European immigration and northeastern money, the city became an economic powerhouse that dominated the Upper Midwest and colonized the Far West. Within a decade, as St. Louis became one of the nation's most important cities, politics—like everyday life—acquired a more complex character. On the one hand, the city's growth was so commanding that anything seemed possible for the future. On the other hand, that growth came with serious costs. Within the city, social disorder ruled the day; without, the Missouri hinterland grew socially and politically fearful of the growing metropolis. St. Louis helped to bring prosperity to the rural areas, but its economic dominance fostered resentment, as did its strange immigrant ways—giving birth to an enduring St. Louis-outstate split. St. Louis's staunch unionism during the Civil War ensured Missouri's loyalty to the Union as well, but the dislocations of the sectional crisis had already permanently damaged the city. Inflated rhetoric about St. Louis's destiny would, in fact, remain the common coin of the city's boosters long after it was devoid of any reality. It was in the 1870s, when some of their claims of St. Louis's immutable path to preeminence were shrillest, that many sensible people began to realize that the coming glory had passed the city by.

What follow are a series of biographical sketches of representative men who dominated St. Louis from its founding into the Gilded Age—a period that frames the arrival, career, and passing of Henry Shaw. All were national figures. While one or another attained temporary ascendancy, there were no permanent victories; all of these men failed. For all of their fierce effort, world-making—at least the making of a world of their choosing—lay beyond their grasp. Yet their power and influence were undeniable. If the village witnessed by Shaw on his arrival in 1819 was a promising but still unformed place, this group's struggles were essential elements in transforming St. Louis, by the close of the nineteenth century, into a city that today's citizens would recognize as their own.[1]

Prologue *Auguste and Other Chouteaus*

In 1764 St. Louis's first booster, Pierre Laclede, declared, "I have found a situation where I am going to form a settlement which might become hereafter one of the finest cities of America."[2] Laclede had been brought into a partnership that had received from the French government in New Orleans the monopolistic right to trade with the Indians along the Missouri River and the west bank

---

1. I would like to acknowledge the assistance of Christyn Elley, a historian at the Missouri State Archives, in the preparation of this essay.
2. Auguste Chouteau, "Narrative of the Settlement of St. Louis," in *The Early Histories of St. Louis,* ed. John Francis McDermott (St. Louis: St. Louis Historical Documents Foundation, 1952), 48.

of the Mississippi River. Fourteen years later Laclede died in debt, broken in spirit. But his settlement succeeded. Despite his personal financial troubles, the trading post he founded was soon firmly grounded. Many French settlers living on the east side of the Mississippi migrated to St. Louis at the conclusion of the Seven Years War, when France divided its North American holdings between England and Spain. Once St. Louis became Spain's administrative center for Upper Louisiana, its stability and long-term importance were assured.

With Laclede's death, leadership of the little village passed to his stepson, Auguste Chouteau, who in his teens had overseen the creation of the original settlement.[3] Until after the War of 1812 a half-century later, St. Louis was dominated by the Chouteau family. Auguste, the family patriarch, was of particular importance, although his younger half-brother, Pierre, played a significant role as well. Still, their power did not just fall from heaven. As the town's founders, they began in an advantageous position, but they kept that position for fifty years because they were hardworking and crafty. Few could resist the Chouteau charm when they chose to show it. But the family could be ruthless with rivals, and they had a remarkable ability to so thoroughly intertwine public business with their private endeavors as to give them monopolistic powers, particularly over the fur trade.

Chouteau charm was, in fact, frequently required. Shortly after the Laclede-Chouteau family received the French colonial government's concession, in 1764, they discovered that France had given Louisiana to Spain. Accordingly, they fell to wooing the series of Spanish lieutenant governors who came to rule Upper Louisiana during the late eighteenth century. In this they were very successful, despite the occasional vow of some temporary Spanish official to re-order power relationships. The Spanish governors had, in fact, few resources with which to rule. The Chouteaus, in turn, put their more ample resources into the service of state policy, and special trading concessions and large Spanish land grants followed. The deals all went down as nicely as the Chouteau family's superior wine.

When Napoleon reacquired Louisiana for France in 1800, only to turn it over to the Americans three years later, the Chouteaus were stuck with troublesome Yankee rulers, but generally the old formula of mutual, public-private compromise worked. William Henry Harrison found this arrangement well enough to his liking during his brief tenure as territorial governor; the next governor, the double-crossing Aaron Burr confederate General James Wilkinson, found

---

3. William E. Foley and C. David Rice, *The First Chouteaus: River Barons of Early St. Louis* (Urbana: University of Illinois Press, 1983). For more on Laclede, the Chouteaus, and the other figures discussed in this essay, see Lawrence O. Christensen et al., eds., *Dictionary of Missouri Biography* (Columbia: University of Missouri Press, 1999).

it quite easy as well, as did the explorer, and later territorial governor, William Clark.

Yet for all their power and influence, the Chouteaus had little civic vision. Despite the high political contest that later followed, the family had no grander aspiration than to make money from the Indian fur trade. As the preeminent men of the village, indeed of Upper Louisiana, the Chouteaus had a self-conceived role to play in the community that involved civic obligations. Their good works, however, were mostly, if not entirely, self-interested—the sort that, in modern terms, could be described as creating a good business climate. One cannot dismiss the Chouteaus as simply selfish; they were just businessmen, and reasonably enlightened ones, as they conceived their role.

In the years after the Louisiana Purchase, St. Louis grew, and so did the number of Chouteau political opponents. The Chouteaus had developed the necessary skills to woo politicians one by one, but not in larger political groups. Many newcomers challenged the vast Spanish land grants claimed by the French elite; they believed a good number of these so-called grants were fraudulent. Some undoubtedly were. In the years following the War of 1812, the French elite began to lose their power and the new Americans, some initially tied to the French, began to take over.[4] Unlike the Chouteaus, however, these were men of broad public vision, and although St. Louis would have more than its share of venal politicians, the best of these men cared more for glory than money. Among the many examples of this group, the most important is Thomas Hart Benton.

Thomas Hart Benton and the Yeoman West

Senator Thomas Hart Benton dominated Missouri politics during the first half of the nineteenth century. Indeed, only Harry Truman surpasses him in significance as a national political figure from the state. Benton won five successive terms as United States senator between 1820 and 1850, and continued to play a significant role during the 1850s as Missouri followed the nation into civil war. Benton was something of a paradox: a sincere champion of the farmer who lived as a lawyer and newspaper editor in St. Louis, having found farming repugnant to his nature. Though he made common cause with Missouri's gentry, he was most devoted to the aspirations of Missouri's poorest farmers. Yet even as he addressed them at endless political rallies, he never sought to do so as an equal. Benton may have spoken for the ill-clad, homespun people, but he did so in

---

4. Stuart Banner, "Transition," chap. 5 in *Legal Systems in Conflict: Property and Sovereignty in Missouri, 1750–1860* (Norman: University of Oklahoma Press, 2000).

Thomas Hart Benton (photograph, c. 1856). Benton, champion of western agriculture as well as of St. Louis commercial interests, did as much as any single public figure to position Missouri prominently on the broad sectional issues that faced the expanding nation in the decades leading up to the Civil War. *Missouri Historical Society*

sartorial, almost dandified, dress, for he regarded himself very self-consciously as a refined gentleman. And although on occasion he could be quite crude, this champion of the ill-educated impressed and sometimes tormented his fellow senators with his classical learning and obsessive research, which found form in speeches of forceful bombast. On one occasion, a man allegedly approached Benton and said, "Senator, this boy walked two hundred miles to hear you." Benton turned to the boy and said, "Young man, you did right."[5] While many

---

5. Quoted in William Nesbit Chambers, *Old Bullion Benton: Senator from the New West* (Boston: Little, Brown, 1956), 338.

believed there was a planned Democratic presidential succession that was to run from Jackson to Van Buren to Benton, it was all derailed by the financial panic of 1837 and the rise of the Whigs. Benton, however, knew he excelled as an advocate, not as an administrator, and he routinely turned aside the common talk of a presidential candidacy. Even so, he was a consummate political insider. Despite his fervent claims that he served as the voice not only of Missouri, but of the entire West, Benton lived primarily in Washington after his Senate career was firmly established.

Thomas Hart Benton was born into a prosperous North Carolina gentry family in 1782.[6] The early death of his overextended father left the family in financial difficulties, and family hopes settled on the obviously talented Thomas. Benton, however, disgraced himself at the University of North Carolina when he was expelled for pilfering small amounts of cash from his roommates. This episode would be recounted throughout his life by political enemies, which made him very touchy about his honor. Not long after he left the university, his family's financial circumstance dictated that it remove to lands his father had speculated on in Tennessee. After a brief but distasteful experience helping work the land, Benton found his way into law and then into politics. He won election to the Tennessee Senate in 1809, but temporarily dropped formal politics to seek military glory during the War of 1812. Initially a favorite of General Andrew Jackson, his brother involved him in a quarrel with one of Jackson's close friends. The eventual result was a nasty brawl in Nashville in which Benton shot Jackson in the arm, while he, himself, received a saber slash. Remarkably, no one was killed. At the brawl's conclusion, Benton dramatically broke Jackson's sword in two before an astonished crowd. When Jackson subsequently marched off to glory at New Orleans, Benton was left behind as a recruiting officer. As one historian notes, "Shooting Jackson in Jackson land was like knifing the Pope in the Vatican."[7] His Tennessee career was over.

In 1815, at age thirty-three, Benton joined the vast middle-border migration heading into Missouri following the war. If he had evinced sympathy for the poor farmer in Tennessee, in St. Louis he made fast friends with the French and American elite. Their principal interests consisted of promoting their international fur trading operations and protecting their large Spanish land grant claims, which the United States government sometimes suspected had been inflated by fraud. Benton made his stake as their advocate at the bar.

---

6. Benton's rise is chronicled in Chambers, *Old Bullion Benton,* and Elbert B. Smith, *Magnificent Missourian: The Life of Thomas Hart Benton* (Philadelphia: J. B. Lippincott, 1958).

7. William W. Freehling, *The Road to Disunion: Secessionists at Bay, 1776–1854* (New York: Oxford University Press, 1990), 541.

In 1818 he became the editor of the *St. Louis Enquirer*, which he intended to be principally a vehicle for his political aspirations. With the prospect of Missouri statehood likely, Benton became its foremost proponent. Statehood did not come easily. Opponents to Missouri's request to form a state government precipitated a national crisis as slavery's foes sought to prevent the spread of the institution to the vast territory acquired through the Louisiana Purchase. This obstructionism infuriated the promoters of Missouri statehood, especially Benton, who unreservedly declared his support for slavery and styled his polemical arguments as resistance to federal dictation. Eventually, the so-called "Missouri Compromise" granted Missouri statehood in exchange for a law forbidding further slavery in the Louisiana Purchase area north of Missouri's southern border and balancing Missouri's entry into the Union as a slave state with the entry of the free state of Maine. In Missouri's first elections in anticipation of imminent statehood, Benton watched in surprise as his French-Anglo allies, led by territorial governor William Clark, went down to inglorious defeat. He quickly became aware of his own tenuous political circumstance. He had been hoping for a United States Senate seat. In the end, his prominence in the campaign for statehood sufficiently counterbalanced his unpopular friends, and all proved well. When the first legislature met, it narrowly chose him as Missouri's junior senator.

Benton learned from his scare, and soon became the Senate's most conspicuous spokesman for the ambitions of common farmers. More specifically, he spoke as the voice of the "New West," which he believed would one day succeed the East as the dominant section of America, as it both grew and shaped the development of what became known as the Far West. Benton had a vision of a sprawling land of virtuous independent yeoman farmers, whom he considered the backbone of the republic. Making this vision a reality was his guiding light and he fought fiercely against anything he thought stood in its way. First, farmland had to be easy to acquire. While the national government took in much of its revenue from the sale of public land to settlers, land prices could not be an impediment to western settlement. Benton, with some success, urged that prices be lowered, but he never achieved his goal of allowing unsold land to eventually be freely given to any settler—something that did not occur until Congress passed the Homestead Act during the Civil War. He also championed hard money—that is, gold and silver coin as the basis for the money supply—believing that unstable and sometimes worthless paper money defrauded both city laborers and country farmers. He thus earned the popular sobriquet "Old Bullion Benton."

When Andrew Jackson dramatically assumed the presidency in 1828, events turned Benton's way. If history remembers Clay, Calhoun, and Webster better than Benton, it is probably because he was not fighting Jackson as were the other three. Remarkably, the two men made up and Benton became Jackson's principal ally in the Senate. During Jackson's two administrations, Benton fought not only for western expansion and agrarian and monetary reforms, but also against monopolies and legislated special privileges that seemed to stand in the way of the common man, and he served as Jackson's point man in the famous crusade against the Bank of the United States. Benton's influence continued to wax strong during the presidency of his friend Martin Van Buren. From the late 1820s to the early 1840s, Benton and Jacksonian democracy were at their height.

As events unfolded, Benton the urban-dwelling farm advocate felt no distinction between St. Louis and the empire that would unfold to its west. They lived in organic unity. Yet in a rapidly developing nation, with little authority to control its wild swings of boom and bust, Benton did his best to undermine the institutions and policies, like the Bank of the United States and a more orderly land policy, that might have restrained the economic and social anarchy that ensued. With the advent of the market economy and of migratory and immigrant flows, the old social order inherited from colonial days collapsed. Benton himself helped to accelerate that change, even as he looked to spread an older agrarian ideal. In so doing, he unwittingly helped divide St. Louis from its hinterland politically, eventually destroying his own career in the process. There were other St. Louisans, however, who tried to promote greater social stability. They agreed with Benton that the rapid development of St. Louis and the West was highly desirable, but hoped the region's coming prominence would shape the character of the nation in yet another way. Two of these were Edward Bates and William Greenleaf Eliot.

Edward Bates: Legal Whiggery

To be a political opponent of Thomas Hart Benton in antebellum Missouri gave one an opportunity to watch politics from the sidelines. It did not begin that way for Edward Bates, however. Bates was born in prosperous circumstances in Belmont, Virginia, in 1793.[8] Like Benton, Bates lost his father early, when he was twelve. Fortunately, a number of well-to-do relatives came to his aid and the bookish boy received a good education. Like Benton, he also volunteered for service during the War of 1812, attaining the rank of sergeant

---

8. Marvin R. Cain, *Lincoln's Attorney General: Edward Bates of Missouri* (Columbia: University of Missouri Press, 1965).

Edward W. Bates (steel engraving by F. Garsch for *DeBow's Review*, c. 1855). Bates, a practicing attorney who also served in a wide variety of state and federal positions during his career, owed his political longevity in St. Louis to a relatively low public profile and a conservative, Whig point of view that persisted throughout a period of controversy and change. *Missouri Historical Society*

by the end of his service. From the army, Bates came to St. Louis, where his brother, Frederick, had established himself as a frontier politician. (Frederick Bates later became Missouri's second governor.) Edward studied law with Rufus Easton, one of the foremost lawyer/politicians of St. Louis, and, with the help of his brother and Easton, soon found clients among St. Louis's wealthy elite. In 1818, Territorial Governor William Clark appointed him attorney for Missouri's northern district.

In 1820 Bates was elected to the Missouri constitutional convention. During its proceedings he played a prominent role in designing a strong independent judiciary, which he regarded as a necessary counterweight to the inherent instability of popular government. He had little confidence in the wisdom of common men, the very citizens who were the bedrock of Benton's political power. Bates accepted the concept of white-manhood suffrage, but he wanted property qualifications for both voting and office holding. Yet if Bates had little faith in people, he had a great deal of faith in the law. He loved it, cherished it, and saw it as an instrument of order in a disorderly world. He regarded the subsequent popular assaults launched against the judiciary during the Jacksonian era as an attempt to desecrate the temple of justice. Bates was a staunch supporter of western economic growth as guided by Henry Clay's "American System" of high tariffs, internal improvements, and support of the Bank of the United States. He believed that growth and social change were good things, but that they had to be carefully calibrated and disciplined. He had little taste for the political anarchy and land giveaways promised by Benton.

Under Missouri's first constitution, the governor appointed cabinet officials, and Governor Alexander McNair made Bates the state's first attorney general.[9] Bates's law partner, Joshua Barton, received McNair's appointment as the first secretary of state. Joshua's older brother, David, had been elected Missouri's senior senator to Benton's junior position in 1820. David Barton had been Missouri's most popular politician of the late territorial period, but political lines had muddied. In 1824 the avid Jacksonian-to-be, Benton, supported his wife's cousin, Whig-to-be Henry Clay, for president; Barton supported New Englander John Quincy Adams. Barton and Benton were soon at war. In a surprise, Andrew Jackson won the presidential popular vote in 1824 but, thanks to an alliance between Adams and Clay, lost the office when the failure of any candidate to achieve an outright majority threw the contest into the House of Representatives. In Missouri, with Benton's support, Jackson's subsequent presidential campaign in 1828 took on the fervor of a cause. Barton's espousal of Adams's reelection left him well out of step with his fellow citizens. Jackson and Benton's crushing victory at the polls left their political opponents prostrate before them.

Bates's close ties to the Barton faction spelled his political doom. Earlier, in 1822, he had won election to the Missouri legislature, and not long thereafter to a single term in the United States Congress. With the destruction of the Barton wing of the old Jeffersonian party, Bates was defeated for reelection. He subsequently moved to St. Charles County to practice law and became a gentleman farmer. Later he served single terms in the Missouri House and Senate,

---

9. Missouri's first executive government was a strictly St. Louis affair: the first governor, lieutenant governor, secretary of state, state auditor, state treasurer, and attorney general were all from St. Louis. St. Louisans would continue to dominate state offices throughout the 1820s.

but he eventually gave up electoral politics. While others acknowledged him as Missouri's foremost Whig politician, he shunned the limelight, preferring to counsel others more willing to serve as cannon fodder for Benton's political machine. By the early 1830s, Bates's day had come and gone, but it was to come again.

William Greenleaf Eliot: New Englandizing the West

William Greenleaf Eliot would undoubtedly have been annoyed to be lumped in with politicians, even though he found their company congenial enough when he wanted something done. His mission was that of Jesus Christ.

Eliot was born in New Bedford, Massachusetts, in 1811. He grew up in Washington, D.C., where his father had an important position with the United States Post Office, and there attended Baptist Columbian College (now George Washington University). Upon graduation he went to Harvard Divinity School and was ordained as a Unitarian minister in 1834. Eliot's time at Harvard coincided with a flowering of New England culture, marked most conspicuously by the transcendentalist movement. Eliot dabbled in the shallows of the movement, teaching himself German and reading idealist philosophy, but the contemplative life did not really suit his temperament. Many years later, he told the story of walking across Boston Common with the self-consciously brilliant Margaret Fuller, who was to win fame as a transcendentalist writer. With a sigh, Eliot told her he sadly regarded his intellect as only second-rate—fit at best to communicate the thoughts of first-rate minds to others. Eliot, who hoped to elicit contradiction from Fuller, was shamed when she brightly informed him how refreshing it was to run across someone with such an accurate understanding of his own abilities.[10]

But Eliot *was* brilliant. He had superior organizational talents, but his genius was for inspiring people to do what he wanted them to do—and this, remarkably enough, through an appeal to the better angels of their nature. Unitarianism in the early 1830s was going through its own version of the Second Great Awakening of which transcendentalism was a part. Although never a transcendentalist, Eliot felt the evangelical fever and supplemented his divinity school learning with good works such as holding Sunday school for prisoners. Upon his ordination, he decided that it was his duty to bring Unitarianism to the heathen West. This was not a very original notion. Indeed, his slightly older friends

---

10. Earl K. Holt III, *William Greenleaf Eliot: Conservative Radical* (St. Louis: First Unitarian Church of St. Louis, 1985), 21. See also Charlotte C. Eliot, *William Greenleaf Eliot, Minister, Educator, Philanthropist* (Boston and New York: Houghton, Mifflin, 1904).

William Greenleaf Eliot (daguerreotype by Thomas M. Easterly, c. 1850). Eliot came to the western frontier from his native New England armed with a combined missionary and social zeal. From the pulpit of the Unitarian Church of the Messiah and, later, from the newly formed Eliot Seminary (later Washington University), Eliot devoted himself to both causes in fast-growing St. Louis. *Missouri Historical Society*

James Freeman Clarke and William Henry Channing had already left Boston to save Louisville and Cincinnati, respectively.[11] Eliot chose to save St. Louis.

That St. Louis needed saving was beyond doubt—New Englanders had been trying for some time. Salmon Giddings and Timothy Flint, both from Massachusetts, had sought to bring the salvation of Presbyterianism to the so-called "land beyond the Sabbath" in 1816.[12] By the time Eliot arrived in 1834, the Protestant beachhead was well established, but life was still pretty grim for someone used to Boston refinement. The chronically muddy streets, roaming pigs, and casual cruelty to both people and animals were impressive to any cultivated easterner—and then there was human slavery. Channing gave up on Cincinnati with barely a struggle; Clarke managed to say enough injudicious things about slavery as to shorten his welcome in Kentucky, allowing him to happily retreat back to Boston. But Eliot stuck. His Unitarianism was so conservative, and his devotion to Jesus so fervent, that he would be judged a moderate by today's mainline Protestant standards. But such was the reputation of Unitarians in 1834 that when he showed up in St. Louis he attracted a modest amount of attention from those wanting to see what an atheist minister looked like. After the curious departed, Eliot said, he was lucky to get eight parishioners to attend services in good weather. Within a few years, however, he had created the most influential church in the city, the Church of the Messiah, and had become the Johnny Appleseed of Unitarianism for the whole Mississippi River Valley. Charles Dickens took time out of his censure of the United States in his travel book *American Notes* to praise Eliot, and the admiring Ralph Waldo Emerson on a visit to St. Louis denominated him the "Saint of the West."[13]

What distinguished Eliot, however, was not his ability to impress the literati, but his ability to transcend class—and in this, perhaps, he was a transcendentalist after all. Few were as ready to sit with the poor, the criminal, and the wretched as Eliot, but his gift lay in his ability to get the well-to-do to pay for his good works. Eliot fed the poor out of his church, took care of orphans, and was

---

11. Holt, *William Greenleaf Eliot*, 20. See also Octavius Brooks Frothingham, *Memoir of William Henry Channing* (Boston: Houghton Mifflin, 1886), 88–92, 144–150, and James Freeman Clarke, *Autobiography, Diary, and Correspondence*, ed. Edward Everett Hale (1891; reprint, New York: Negro Universities Press, 1968). For a discussion of the larger context of Eliot's decision to move west, see Rush Welter, "The Frontier West as Image of American Society: Conservative Attitudes Before the Civil War," *Mississippi Valley Historical Review* 46 (March 1960): 593–614.

12. Kenneth H. Winn, "Salmon Giddings" and "Timothy Flint," in Christensen et al., eds., *Dictionary of Missouri Biography*, 336–37 and 304–5, respectively. See also William E. Foley, *The Genesis of Missouri: From Wilderness Outpost to Statehood* (Columbia: University of Missouri Press, 1989), 269–82.

13. Charles Dickens, *American Notes for General Circulation* (1842; reprint, New York: St. Martin's Press, 1985), 159; Lorin Cuoco and William H. Gass, *Literary St. Louis: A Guide* (St. Louis: Missouri Historical Society Press, 2000), 39.

the principal guiding spirit behind the creation of the Missouri School for the Blind. He was a careful but avowed abolitionist at a time when the profession of even mild antislavery sentiments was dangerous. He helped create cultural institutions such as the St. Louis Academy of Science, the St. Louis Museum of Fine Arts, and the Missouri Historical Society. He was a friend to Henry Shaw and the Missouri Botanical Garden. During the Civil War he founded the Western Sanitary Commission, which heroically provided assistance to soldiers—wounded, ill, or lonely—political refugees, and Confederate prisoners.

Eliot had his finger in many pies, but education was his main interest. His campaign led to the first city tax support of public schools in 1850. Most important, he founded Washington University (originally incorporated as Eliot Seminary) in 1853 with the financial support of his parishioners.[14] That the university survived had much to do with his brilliance as a fundraiser. In 1870 Eliot resigned from the pulpit and became Washington University's third chancellor. He also founded Smith Academy for boys and Mary Institute for girls. Financial backers named the girls' school after an Eliot daughter who died young. Smith Academy was meant to serve as the feeder prep school for Washington University, but fed St. Louis's wealthy children mostly to Harvard and Yale.

While Eliot's persuasive skills were extraordinary, his message fell on fertile ground. When St. Louis's economy and population exploded in the 1840s and 1850s, the result was highly stimulating. The city was flush with impending greatness, but also a precarious place. Growth was welcomed, but it seemed to many too quick, too dramatic, and too close to utter chaos. Eliot's promise was that he could control that chaos. He could bring people to Jesus and he could mold institutions that would civilize the city. Eliot would have been mortified at the notion that social control was part of his mission. His saw his work as Christ's work, nothing more and nothing less. His parishioners and sympathizers were also sincere, but clearly Eliot hoped to remake St. Louis into a little piece of New England, and if he did not succeed, he nevertheless left an indelible mark on the city.

St. Louis versus Missouri: the Great Divide

In 1999, St. Louis Deputy Mayor Michael Jones made a joke: "What do you get if you take St. Louis and Kansas City out of Missouri?" Answer: "Arkansas." Urban-rural divides are common in America: the split between New York City

---

14. Ralph E. Morrow, *Washington University in St. Louis: A History* (St. Louis: Missouri Historical Society Press, 1996), chaps. 1–5.

and upstate New York, Chicago and downstate Illinois, are well known. Today many St. Louisans will regretfully admit that they are in Missouri but declare they are not *of* it, while rural legislators talk of ceding the city to Illinois. This has not always been the case. Until the late 1840s no great sense of separation existed between the town and the state; even after that, it took a number of decades to mature. St. Louis had, for example, sought to retain its capital status after the approval of statehood; when delegates to the state's constitutional convention determined instead to move the capital to a more central location, they did so without any St. Louis bashing, and the city's representatives complained little about the loss. Outstate prejudice against the town was insignificant.[15]

Still more telling, if informal, evidence rests in the names of important St. Louis businesses and institutions dating from the early and mid-nineteenth century. Today, it is a rare St. Louisan who would name a local concern after the state of Missouri. Yet even within the city's boundaries, the name "Missouri" was once found nearly as often as "St. Louis," a fact attested to by, among other concerns, the Missouri Hotel; the Bank of Missouri; the Missouri Insurance Company; the *Missouri Gazette;* the *Missouri Republican;* the *Missouri Argus;* the *Missouri Saturday News;* the *Missouri Reporter;* the *Missouri Democrat;* the St. Louis Missouri Silk Company; and the Missouri Institute for the Education of the Blind.[16] Interestingly, by the time of the founding of the Missouri Botanical Garden (1859) and the Missouri Historical Society (1866), the naming of institutions after the state had become more rare. By the Civil War's conclusion, it was far more common to name new institutions after the city.

The change that would forge the great distinction between St. Louis and outstate Missouri first appeared clearly in the 1840s when the city's economy and population exploded. This occurred in the face of a cholera epidemic in 1849 that killed more than four thousand, followed swiftly by a devastating fire that destroyed much of downtown. Dynamic urban disorder was the reality of the day. This was William Greenleaf Eliot's fertile ground. Importantly, the white population growth made slave labor insignificant. In 1830, 20 percent of the town population was enslaved. By 1860 the proportion was less than 1 percent.

15. For Jones's joke, see Jerry Berger's column in the *St. Louis Post-Dispatch,* December 5, 1999; for urban-rural divides today, see "Urban vs. Rural Split on Prop. B Triggers Sniping," *Jefferson City News-Tribune,* April 8, 1999. On the lack of urban-rural rivalry in the early nineteenth century, see the following newspapers from 1820: *St. Louis Enquirer,* July 19, October 21, October 28; *St. Louis Missouri Gazette & Public Advertiser,* December 6; and *Franklin Missouri Intelligencer,* April 22, May 13, June 24.

16. The 2000 Southwestern Bell telephone book featured nearly nine pages of businesses named for St. Louis versus less than two pages of those whose name is derived from the state. Moreover, many, perhaps most, of the latter were branches of statewide organizations or had a statewide mission. My conclusion about nineteenth-century St. Louis is based upon a list of geographically named institutions between 1794 and 1888 compiled by Christyn Elley.

There were more free blacks in the city than slaves.[17] Indeed, if St. Louis and Missouri were both originally populated principally by immigrants from the Upper South, large numbers of German immigrants and a smaller, but significant, number of citizens from the Northeast were moving in as well. This new population, as a whole, was not abolitionist-minded, but was hostile to slavery as a labor system. As St. Louis's population ballooned during the 1840s, it became increasingly more diversified economically, and served both as an entrepôt for the Upper Midwest and the gateway to the Far West.[18] In doing so, it came to have far more in common economically with the North than with the South. It took no great analytical skill to see that northern industry and farming were prospering far more than the agrarian South.

But Missouri agriculture *was* prospering. Missouri's most prosperous agricultural region, the Boonslick, was dominated by a sophisticated landed elite.[19] Born in the Upper South, typically Virginia, North Carolina, Kentucky, or Tennessee, most thought of themselves as southern. They were slave-owners, though rarely on the scale of the grandees of the Deep South. Their more common neighbors often owned one or two slaves and worked side by side with them in the fields. Missouri as a whole whitened over the course of the nineteenth century—slaves made up a mere 10 percent of the population, the lowest slave population of any slave state except Delaware. In the Boonslick, slaves remained 30 percent of the population. Despite an important sprinkling of Henry Clay Whigs in central Missouri, the Boonslick landowners formed the heart of the outstate Democratic party. Their leaders were known as the Central Clique, and few obtained higher office without their approval. During the 1820s and 1830s, Thomas Hart Benton spoke for them and their aspirations.

As the 1840s began, a change took place. Boonslick farmers had always been commercial farmers, even when they started out as small landowners. In the years following the land rush into the interior in 1819, they had commonly arranged for their produce to be transported to New Orleans. By the 1840s, their business was with St. Louis and they felt a sense of economic dependency. Often they resented it.[20] St. Louis began to seem an alien culture, in some ways

---

17. Manuscript schedules for the Fifth (1830) and Eighth (1860) Census of the United States, St. Louis County, Missouri State Archives, Jefferson City, Missouri.

18. Jeffrey S. Adler, *Yankee Merchants and the Making of the Urban West: The Rise and Fall of Antebellum St. Louis* (New York: Cambridge University Press, 1991).

19. A good parallel discussion of the politically and socially important Boonslick region during these years is R. Douglas Hurt, *Agriculture and Slavery in Missouri's Little Dixie* (Columbia: University of Missouri Press, 1992).

20. Christopher Phillips, *Missouri's Confederate: Claiborne Fox Jackson and the Creation of Southern Identity in the Border West* (Columbia: University of Missouri Press, 2000), 115–21; Adler, *Yankee Merchants,* 110–44.

literally so, with the inrush of German and Irish immigrants. While as a group landowners had prospered—the struggling farmers of the 1819 land rush had become well-to-do and landownership was widespread—it was clear that the economic dynamo to the east was leaving them behind materially and diminishing their political voice.

Granting that there were people on various sides of complex issues, St. Louis and outstate Missouri struggled for supremacy between 1844 and 1875. Initially, outstate Missouri won this struggle, but once the Civil War began, St. Louis's greater resources, backed by the power of the federal Union, carried the state, and, for a time, the city's politicians attempted to assert its varied visions over the state as a whole. The adoption of the 1875 constitution marked, in part, the uneasy truce that has, more or less, continued to this day.

For nearly a quarter of a century, Thomas Hart Benton successfully rode his coalition of outstate interests and St. Louis interests as one horse. After 1846, he discovered he was actually riding two, and when they diverged, they pulled his political career apart.

Benton among the Ruins

During the 1844 presidential election, cracks began to spread across the broad base of Jacksonian democracy. Benton, a longtime advocate of western expansionism, became deeply troubled over the campaign to annex Texas, which he saw principally as a conspiracy to spread slavery.[21] Benton thought himself a Westerner rather than a Southerner. Although the Civil War has obscured western sectionalism as an antebellum counterpart to Northern and Southern sectionalism, it was a common sentiment at the time. Benton based his national agrarian political strategy on a South-West political alliance. But he was a staunch Unionist. He had been furious with John C. Calhoun during the "nullification crisis" of 1832–1833, when South Carolina attempted to assert the preeminence of state over federal authority. In 1844 he perceived Calhoun and the slave power conspirators at work in the forcible annexation of Texas against the wishes of Mexico. As in the nullification crisis, he believed that this base attempt to spread slavery again threatened to disrupt the union. But annexation and the Mexican War came, and when it did, it was highly popular in Missouri.

21. For the story of the challenge to Benton's power and his eventual overthrow, see—in addition to Chambers, *Old Bullion Benton;* Smith, *Magnificent Missourian;* and McCandless, "Thomas Hart Benton"—Phillips, *Missouri's Confederate,* 164–80, and Clarence H. McClure, *Opposition in Missouri to Thomas Hart Benton* (Warrensburg: Central Missouri State Teachers College, 1926).

As America and St. Louis changed, Benton ran into problems. Although he loyally supported the Mexican War after its declaration, his initial opposition to it and his denunciations of it as a slaveholders' conspiracy badly eroded his once strong support among the Central Clique. Slavery was a minor institution in Missouri and Benton hoped it would become more so. Benton was hardly an abolitionist, but he came to believe that Missouri's greatness depended upon attracting white immigrants—immigrants who would settle in Missouri if that meant not competing against slave labor. He was, above all, a nationalist, who despised disunionism. As much as he disliked militant abolitionists, he feared militant slaveowners more.

By the late 1840s, Missouri's Democratic party began to divide. While Missouri's smallest yeoman farmers continued to support Benton, the powerful Central Clique that dominated the outstate party identified with Southern slave culture, and was in sympathy, if not perfect agreement, with that culture's more aggressive defenders. To them, Benton looked more and more like a traitor to the Southern way of life. Benton won reelection in 1844, but by a much narrower margin than in the past. During the late 1840s, Benton became Calhoun's leading opponent, and his vitriolic attacks on what he saw as disunionism led Missouri's influential pro-Southern Democratic leaders to seek his defeat. In a brutal 1850 campaign, they succeeded by joining to elect St. Louis Whig Henry Geyer, an ardent advocate of slavery, as senator.

St. Louis, by contrast, had become the bastion of "free labor," bolstered by thousands of German immigrants, nearly all of whom hated slavery for economic reasons, and an influential minority of radical intellectuals, who hated it for ideological reasons. Although some members of St. Louis's dying Whiggery went off to flirt with the anti-immigrant American (Nativist) party, many suddenly found new virtue in Benton, a man whose monetary and land policies they had despised for nearly three decades. Time had provided an ironic twist as common enemies made new friends. Although he often did not like it, antebellum America's most forceful western agrarian advocate now fought to align Missouri with the market revolution of the Northeast from an urban power base. Despite the loss of his Senate seat and his sixty-eight years, Benton had no desire to leave politics. In 1852 he won election to Congress as a representative of St. Louis's first district. He would spend much of his declining political career fighting for a transcontinental railroad originating in St. Louis. Yet, in a harbinger of a new world to come, Benton, a relic of Jacksonian America, sat in his last session in Congress next to New York Congressman William Marcy Tweed, the "Boss" of the Gilded Age.

Defeated in his 1854 reelection bid for Congress, Benton ran for senator against the militant proslavery incumbent David Rice Atchison and Whig Alexander Doniphan. The Missouri legislature became so hopelessly deadlocked in deciding the election that the state remained without a second senator for two years. In 1856 Benton ran for governor and lost to the staunch proslavery Democrat Trusten Polk. As fierce an old man as he was a young one, infirm in body but not spirit, he wrote and spoke of the burning political issues of the day and of his past until he died in 1858. He is buried in Bellefontaine Cemetery in St. Louis.

Benton's life stretched from elitist French St. Louis to Jacksonian republican agrarianism to the dawn of industrialism and the coming of the Civil War. For nearly three decades he was a master of his world. But the seductive promise of agrarian republicanism could not hold together long. The development of a national market economy and the strains of maintaining an obsolete and inhumane labor system blew his world apart when it separated his city from his state.

Barn Burners: Blairs and Brown

In his fierce campaign against pro-Southern Democrats, Benton did not work alone. He was joined by the Kentucky sons of his friend and distant relative Francis Blair—and later their cousin, B. Gratz Brown. Francis Blair had been an intimate of Andrew Jackson and remained, through his Maryland home, the consummate Democratic party insider. The Blair sons, Montgomery and Frank, joined Benton's St. Louis legal practice, but principally as a means of supporting their all-consuming political interests. (Montgomery would serve a term as St. Louis mayor, 1842–1843.)[22] The young men who surrounded Benton shared his concern for the small farmer and the city wage earner. They scorned the plantation masters as the ruin of the country. Benton, the Blairs, and Brown had all been slaveholders, but they believed the Southern planters' obsession with slavery had made them crazy. Over time, the issue of the role of blacks in America would divide the younger generation. Frank Blair fiercely disliked Negroes, slave or free, while Brown would eventually champion their political rights. During the 1850s, however, theirs was a white man's movement, designed to liberate white Missourians from competition with slave labor, which they viewed as a bar to the state's prosperity and white immigration.

22. The best biographies of Frank Blair and B. Gratz Brown are William E. Parrish, *Frank Blair: Lincoln's Conservative* (Columbia: University of Missouri Press, 1998), and Norma L. Peterson, *Freedom and Franchise: The Political Career of B. Gratz Brown* (Columbia: University of Missouri Press, 1965). See also Louis S. Gerteis, *Civil War St. Louis* (Lawrence: University of Kansas Press, 2001).

Frank P. Blair Jr. (steel engraving by John Sartain after F. T. L. Boyle, 1862). Blair, although hailing from a well-connected Kentucky slaveholding family, brought with him to St. Louis a fierce opposition to slavery and a desire to keep Missouri in the Union. In the end, his support proved crucial in the contest to keep secessionists from prevailing in St. Louis. *Missouri Historical Society*

Benton wished for a continuation of the conspiracy of silence on slavery that had marked the Jacksonian era. Why bring up things that unnecessarily upset people? Calhoun had much to answer for. The Blairs and Brown obediently followed Benton through the 1840s, but chafed under his direction as time went on. They found his approach too conservative and thought his lifetime in Democratic party service made him too tolerant of party temporizers. The Blairs and Brown wanted action. They were rough and tough men. It was said, "When the Blairs go in for a fight, they go in for a funeral."[23] The three cousins made no bones about being barn-burning "free-soilers"—that is, advocates of keeping slavery out of the western territories.

From their St. Louis base, Frank Blair and Brown agitated against slavery through the *Missouri Democrat,* of which Brown served as the editor. As a member of the Missouri House of Representatives, Brown caused a sensation in 1857 by calling for the abolition of slavery in Missouri. Far from attacking slavery on humane grounds, he typed himself as the defender of free-white labor and called slavery a barrier to economic progress. Much of his analysis began on these old Bentonite grounds, but he did not go on to damn abolitionism and secessionism as twin evils. His speech was designed to work on several levels: the call for slavery's end was designed to appeal to German immigrants, most of whom were hostile to the institution; the call for abolition in the name of economic progress was, in part, designed to appeal to former moderate Whigs who were looking for a home; and the emphasis on free labor was designed to appeal to the aspirations of ordinary farmers who had made up the backbone of Benton's followers. The result, however, was to send his career into temporary eclipse. It brought him a hard nucleus of followers, but it also brought him bitter enemies. Benton thought the speech a rank betrayal, and Brown soon lost office. An unrelated quarrel with his cousin Frank ended with his being forced out as editor of the *Missouri Democrat,* and the two broke off relations as well.

In 1856 Frank Blair, over Benton's wishes, had supported Benton's son-in-law, John C. Fremont, as the first Republican nominee for president, and that same year Blair had won election to Congress as a free-soiler. As the Southern states began to secede in the wake of Abraham Lincoln's 1860 election, Blair took the lead in Unionist efforts to save Missouri for the Union. With the assistance of his brother, Montgomery, whom Lincoln had put in his cabinet as Postmaster General, Blair got William S. Harney, the local military commander, removed; Blair deemed him insufficiently ardent in his Unionism. He was replaced by New England abolitionist Nathaniel Lyon, a regular army captain (soon to be general) who had seen action in "Bleeding Kansas."

23. James Neal Primm, *Lion of the Valley: St. Louis, Missouri, 1764–1980,* 3d ed. (St. Louis: Missouri Historical Society Press, 1998), 244.

In May 1861 Blair and Lyon orchestrated the capture of Camp Jackson, a legally organized militia encampment created by Missouri Governor Claiborne Fox Jackson.[24] Blair and Lyon believed, probably accurately, that the secessionist-minded governor had called for the encampment at Lindell's Grove, on the western edge of St. Louis, with the design of capturing the federal arsenal, on the city's South Side. Unionist precautions had made the capture of the arsenal impossible by the time of the encampment, but Lyon and Blair forced the militia's surrender anyway. Marching the captured militia forces into the city, the federal volunteers found themselves amidst a jeering, hostile crowd. A shot rang out and mortally wounded a federal officer. A riot erupted, and the troops fired on the unarmed crowd, killing twenty-eight and wounding many others. Three weeks after the surrender of Fort Sumter, the Civil War had begun in Missouri. Union forces, composed of St. Louisans, were soon on their way to Jefferson City, putting the governor to flight. Lyon, for his part, would meet a martyr's death in the second major battle of the Civil War, at Wilson's Creek in southwest Missouri.

Henry Boernstein and the Germanic Hordes

The riot in the wake of the capture of Camp Jackson was, in part, an ethnic clash in which predominantly German troops marched the state militiamen through the city to the taunt of anti-immigrant slurs by a gathering crowd. In the end, Germans would prove key to saving Missouri for the Union.[25]

The Napoleonic Wars had left vast populations across the German states economically distressed and politically discontented. By the 1830s, tens of thousands of German immigrants had made their way to Missouri. Most became farmers in Missouri River Valley counties between St. Louis and Jefferson City, but at least five thousand German immigrants, fully a third of the city's 1840 population, settled in St. Louis. Within five years that figure had doubled to ten thousand, helping to elevate St. Louis to the rank of eighth largest city in the

---

24. The best account of the Camp Jackson affair and the origins of the Civil War in Missouri is Gerteis, *Civil War St. Louis*, 97–161. See also Christopher Phillips, *Damned Yankee: The Life of General Nathaniel Lyon* (Columbia: University of Missouri Press, 1990), and William Garret Piston and Richard Hatcher III, *Wilson's Creek: The Second Battle of the Civil War and the Men Who Fought It* (Chapel Hill: University of North Carolina Press), 24–43. William W. Freehling offers a larger context for these events in *The South vs. the South: How Anti-Confederate Southerners Shaped the Course of the Civil War* (New York: Oxford University Press, 2001).

25. *Germans for a Free Missouri: Translations from the St. Louis Radical Press, 1857–1862*, selected and translated by Steven Rowan, with introduction and commentary by James Neal Primm (Columbia: University of Missouri Press, 1983). See also Walter Kamphoefner's essay on immigration in this volume.

United States. By 1860 the city's population had reached 161,000 souls, nearly a third (50,100) of whom were German immigrants and their children.[26]

In the two decades before the Civil War, a common discontent drove one and a half million German immigrants to America, but beyond a shared desire to leave their homeland they remained divided by class, politics, religion, and historical experience. Their principal interest in obtaining New World homes was economic success.

Across Germany, as across much of Europe, democratic revolutions rose and fell in 1848. Many involved in these revolutions came to America in the wake of their failure. Quite often, the ordinary German immigrant who preceded them found the intellectuals and political activists who dominated this group a tiresome lot, full of impractical utopian schemes and ideological blather. Many were staunchly anticlerical, while the great masses of German immigrants were sincere, frequently conservative, Christians. The (often young) revolutionists, in turn, despaired of the ordinary German immigrant as a degraded, uncultured drudge, whose debased existence consisted of nothing beyond a narrow rooting for money.

If mutual suspicions never fully evaporated, events during the course of the 1850s drove ordinary German immigrants into the arms of the 1848 revolutionists. The sheer number of immigrants (which included large numbers of Irish immigrants as well) startled and frightened the older Anglo-American community, and cultural differences heightened those fears. Friction between "native" and immigrant communities erupted into rioting in the mid-1850s. Whatever their flaws, the "Forty-eighters" were skilled at writing, speaking, and political organizing, talents the besieged German community came to value for their defense.

During these tumultuous years, Henry (Heinrich) Boernstein was one of the most important leaders of the St. Louis German-speaking community, and indeed, one of the most important spokesmen for German immigrants across the American Midwest. The son of a liquor and perfume factory manager, Boernstein was born in 1805 in Hamburg and raised in Galizia, near today's Poland–Ukraine border. As a young man, he lived a peripatetic existence as a soldier, journalist, medical student, theatrical director, and actor. By the 1840s he was living in bohemian Paris, where he combined acting with political journalism, and at one point he had a fleeting collaboration with Karl Marx.

In 1849, Boernstein decided he had had his fill of Europe and European politics and joined the hundreds of thousands of German-speaking emigrants who came to the United States at midcentury. Rejecting romantic notions about

---

26. Rowan and Primm, eds., *Germans for a Free Society*, 4.

farming life, he settled in Highland, Illinois, where he set himself up as a hydrotherapeutic physician. Little more than a year later, however, a twist of fate made him the editor of the *Anzeiger des Westens* in nearby St. Louis. Through indefatigable energy he soon turned the paper into one of the leading German-language newspapers in the country and, for a time, the principal voice of the German community in the West. Although exhilarated by his role in the German community, he despaired of the sorry condition of German culture, which he thought languished in a state of brutish moneygrubbing. He responded to this condition with a strenuous effort at cultural uplift through the pages of his newspaper; he created a German-language theater (in which he occasionally acted), and he sponsored and participated in educational efforts initiated by others.

Throughout the 1850s, Boernstein relentlessly championed anticlericalism. In 1851, he published the sensational anti-Jesuit novel *The Mysteries of St. Louis,* which he dedicated to Thomas Hart Benton. He opposed the rising tide of Know-Nothing nativism, and denounced slavery as immoral and inimical to economic progress. In the growing sectional crisis, he prominently allied himself with conservative Republicans as represented by Frank Blair. When war came, he championed the Union cause in the German community and, despite his age, accepted a commission in the federal army, serving for a time as the military commander at the state capital in Jefferson City.

Although Boernstein participated in the European upheavals of 1848, he was not properly a Forty-eighter. Indeed, despite many commonalities with them, he came to regard Forty-eighters as irritating ideologues who advocated unrealistic politics. They, in turn, judged him an unprincipled self-promoter. When Boernstein accepted the post of U.S. consul in Bremen in 1863, his political opponents in Missouri effectively undercut his authority and soon absorbed his newspaper. Boernstein left the consular service in 1866 but decided to remain in Europe, eventually settling in Vienna. Continuing to write for American German-language newspapers, he serialized his memoirs, which were published in book form in 1881.[27] He died in 1892.

Upon their arrival in Missouri, German immigrants attached themselves to Thomas Hart Benton and the Democratic party, attracted by Benton's profession of egalitarianism and his rejection of the nativism that contaminated large portions of the Whig party. As Frank Blair and B. Gratz Brown rose to take Benton's place in the late 1850s, the state's German population followed them into

---

27. Henry Boernstein, *Memoirs of a Nobody: The Missouri Years of an Austrian Radical, 1849–1866,* ed. and trans. Steven Rowan (St. Louis: Missouri Historical Society Press, 1997).

the new Republican party. Along the way, the Germans had to tolerate political intercourse with former Whig nativists and prohibitionists, but in the end the transition proved fairly easy. Ordinary Germans felt economically threatened by competition from the even poorer Irish immigrants, who had also entered Democratic party folds. Moreover, events showed that slave masters dominated the Democratic party. For many Germans, especially the radical elite, it was easy to imagine lordly slaveholders as landed Old World aristocrats. They tied the agrarian republicanism of the American founders to their aspiration for immigrant farmers. America was not only the last best hope for mankind, but in the grand scheme of world history it was the last best hope for the redemption of Germany.

Edward Bates Redux

If German immigrants moved into the Republican party with relative ease, they found some of the party's subsequent history hard to swallow. Key to that eventual disaffection was the newly resurrected political career of Edward Bates.[28] The mounting troubles of Benton's career in the late 1840s coincided with the revival of that of Bates. In 1847 he was chosen to serve as the president of the national Chicago River and Harbor Improvement Convention. His unexpected eloquence on western expansion and economic development brought him national prominence. In 1850 Millard Fillmore offered him the position of secretary of war, which he declined, but others across the country began to solicit his opinions. As Benton proved a die-hard Democrat, even as his party divided, Bates proved a die-hard Whig, presiding over the last national Whig convention in 1856. After a brief flirtation with the American or Know-Nothing party, during which he gave voice to anti-immigrant sentiments (sentiments that would come back to haunt him), he joined the new Republican party. There he found a warm welcome from the acolytes of his old adversary, Benton. Blair and Brown needed Bates. There was hardly a shred of support for the Republican party in Missouri outside of St. Louis. Across Missouri, Republicans were widely regarded as wild-eyed fanatics, "Black Republicans." Bates's well-known conservatism was useful to a party trying to establish itself.

As the 1860 election approached, Bates emerged as a major contender for the presidential nomination, certainly as strong as Abraham Lincoln. Frank and Montgomery Blair and Gratz Brown boosted Bates's candidacy as the key to winning the border states as well as the North, thereby saving the Union.

28. Cain, "Requiem for Whiggery," chap. 3 in *Lincoln's Attorney General*.

Whatever chances he might have had were undermined when key German-immigrant politicians made it clear that they could never support him in view of his espousal of Know-Nothing sentiments during the late 1850s.

Given Bates's importance as a border politician and a rival, Lincoln appointed him the nation's attorney general in 1861. A conservative during a revolutionary period, he regarded the operation of the civil law as sacrosanct. His views put him at constant odds with his more radical colleagues such as Salmon Chase and Edwin Stanton, as well as with the military, all of whom pushed for extraordinary measures to match the extraordinary times. Although he backed the Emancipation Proclamation, he did so because he feared worse, and fervently wished it tied to a colonization-deportation scheme. As a border conservative, he had difficulty framing the destruction of slavery outside of his concern for property rights. Taking citizens' property with no compensation was anathema to him. For the slaves themselves he thought little.

Even though he spent the war in Washington, Bates kept very much involved with events back in Missouri, where factional infighting had grown fierce among Unionists. Unfortunately, near the war's beginning, Bates and the Blairs split over strategy. Bates and his brother-in-law Hamilton Gamble, a St. Louis lawyer who became Missouri's wartime governor, sought to conciliate those who were neutral, or at least not actively pro-Southern, and to use the local militia to keep order. The Blairs demanded the defeat of disunionists by means of a vigorous military campaign with federal troops. Still more politically strident were the growing numbers of so-called "Radicals," led by St. Louis lawyer Charles Drake, who wished to use the war to remake Missouri society. Resigning as attorney general in 1864, Bates returned to fight the Radicals, but his health declined as the Radicals' power waxed. For Bates, furious and ailing, it was too much. He had become an anachronism. Fighting until the end, he died in early 1869.

Ducks and Drakes

Charles Drake, the principal proponent of the Radical movement so despised by Edward Bates, was born in Cincinnati in 1811, the son of nationally famous physician Daniel Drake.[29] After a cruel, if well-meant, upbringing at the hands of his tyrannical father, he drifted through school and a stint in the navy, at last landing in an uncle's Cincinnati law office. He won admission to the bar

---

29. For Drake's life, see David DeArmond March, "The Life and Times of Charles Daniel Drake" (Ph.D. diss., University of Missouri, 1949).

in 1833. His indifferent prospects in Cincinnati led him west, and he settled in St. Louis. His father's prominence won him easy entrée into St. Louis society and he soon met and married Mary Ella Taylor Blow, the daughter of Captain Peter Blow (who, as chance would have it, was Dred Scott's first master). A good marriage into a well-connected family was matched with a partnership with well-connected barrister Wilson Primm. Drake also formed an intimate friendship with Hamilton Gamble, the future Missouri Supreme Court justice, wartime governor, and, as it turned out, Drake's future nemesis.

As a lawyer, Drake specialized as a debt collector for eastern interests, but his contentious manner, his unpopular field of expertise, and the financial panic of 1837 conspired to undermine his practice. As St. Louis boomed in the 1840s he began to struggle financially; within a short span of years, his wife and two children died. Drake remarried and tried to recoup his fortunes as a Whig activist, but found patronage plums going to others. After an embarrassing retreat to his father's home in Cincinnati, he found a position in Washington as the treasurer of the Presbyterian Board of Foreign Missions.

Drake returned to St. Louis in 1850, temporarily eschewed politics, and rededicated himself to his law practice. By the end of the decade, however, he had reembraced politics, now as a Democrat. In 1859 he won a term to the Missouri legislature, where he offended his colleagues with his condescending manners and offended German immigrants by mocking their Sabbath-breaking beer drinking and antislavery predilections—the "Red Republicanism of Europe," he said, had become "the Black Republicanism of Missouri."[30] He wisely did not stand for reelection in 1862.

From this unpromising situation, Drake underwent a transformation that took him from militant antislavery opponent to militant antislavery proponent. During his term in the legislature, politics had radicalized and emancipation sentiment, which had been unthinkable to all but a political fringe only a few years before, was now widespread. Political divisions became predicated on how, not whether, emancipation should take place. The "conservative" Unionists, who under the leadership of Provisional Governor Gamble controlled Missouri, wished for a gradual emancipation, hemmed in with conditions. The Radicals, as they came to be known, demanded immediate and unconditional emancipation. While Drake did not initially make his sentiments about immediate emancipation clear—and he may have had some reservations—he had no reservations about his hatred for slaveholders, and in 1861 his well-known stance was sufficient to win him a place in the state convention that governed Missouri during the war.

30. March, "Life and Times," 69.

The Radicals roiled in frustration at the Gamble administration's tepid challenge to those considered to hold pro-Southern sympathies, as well as at its cautious support for an emancipation law that would not free slaves until July 4, 1870 (and then with conditions tied to age and employment). On September 1, 1863, the Radical Unionists held a convention in Jefferson City to create a formal statewide organization. Drake served as the convention's keynote speaker. Upon adjournment, a "Committee of Seventy," led by Drake, left for Washington to personally inform President Lincoln of Missouri's troubles and the Gamble administration's near treasonous toleration of those opposing the Union. Lincoln received them courteously, but, to the Radicals' chagrin, the meeting produced nothing, earning the president their enmity. Other setbacks followed as Missouri Supreme Court elections brought Conservatives to power and the Radicals had to swallow Lincoln's renomination. In a more personal setback, Drake watched Radical Thomas Fletcher receive the gubernatorial nomination for which he himself longed.

However, things soon looked up. The Radicals swept Missouri in the 1864 elections. Governor Fletcher, as planned, called for a new constitutional convention, which met at St. Louis's Mercantile Library beginning on January 6, 1865. It was presumed that the convention would pass an amendment to the state constitution that would immediately end slavery, and then adjourn. Drake had different ideas. He came to the convention fully prepared, and wooed a Radical majority into adopting a sweeping new constitution that a loyalty-tested, purged electorate narrowly passed.

The result, which became known as "Drake's Constitution," was popularly characterized by its opponents as a "Draconian code." During the course of the war, Radicalism's power base shifted from St. Louis to rural Missouri, where outstate Unionists had a strong desire to punish their pro-Confederate neighbors, especially in areas where guerrilla activity was strong. They looked to Drake as their leader. The new constitution proscribed thousands of Missourians from voting, holding office, serving on corporate boards, preaching, or acting as a lawyer or teacher because they had at one time expressed a southern sympathy, had relatives who expressed a southern sympathy, or, in practice, sometimes simply because they were Democrats. There were more positive features: Missouri's miserable schools were better financed and augmented by a school system for freedmen.

Many of the most extreme measures did not originate with Drake. He unsuccessfully opposed loyalty oaths for the clergy. (The U.S. Supreme Court later found the oath unconstitutional.) He successfully fought off the confiscation of property belonging to pro-Confederate citizens. While the slaves were freed, in accord with his wishes, they were not enfranchised, although he would soon change his mind on that topic. He likewise turned aside a constitutional proposal to purge the state's judiciary, but he objected only to its placement in the constitution. The Radicals, fearing the courts' ability to undermine their measures, sought to rebuild the judiciary with political loyalists. In March 1865 the new legislature, encouraged by Drake's tireless advocacy, passed the so-called "Ouster Ordinance," declaring more than eight hundred judicial positions vacant as of May 1.

In 1866 Drake worked hard to elect a Radical majority in the Missouri legislature, and when the Radicals triumphed, they rewarded him with a United States Senate seat. The Radicals were the masters of Missouri and Charles Drake was the master of the Radicals. It could not last. The fires that burned bright in the Radical movement were bound to consume it. The first and most devastating challenge came from within the movement. Drake rival Carl Schurz, editor of the St. Louis German-language newspaper *Westliche Post*, ran for Missouri's other Senate seat against a Drake protégé, Benjamin Loan. In a debate, Schurz skillfully taunted Drake into making ethnic slurs against German immigrants, one of the Radicals' core constituencies. The result was not only the end for Loan, but also inevitably for Drake. A new "Liberal Republicanism," as espoused by St. Louisans Schurz and Gratz Brown, would soon take over Missouri and make a grab at taking over the nation as well.

Drake had run out of favor, but not out of patronage. In December, President Ulysses S. Grant appointed him chief justice of the court of claims in Washington, where he served until 1885. He died in 1892 and his cremated remains were interred in Bellefontaine Cemetery.

Blair and Brown: Ascent and Fall

After the war, both Brown and Frank Blair became fierce foes of Reconstruction and political proscription against Confederate veterans. In their struggle with Charles Drake, however, they fared better than their erstwhile partner Edward Bates. Both reached for high office. The Blair family for years had had quiet designs on the presidency for Frank.[31] Early in the war, Blair served as a St. Louis congressman and played a significant role in organizing the war effort. After leaving Congress, he rose to the rank of major general, serving under Sherman and Grant. Sherman, who initially feared him as one of those bothersome political generals, later praised him warmly.

---

31. For these years, see Parrish, *Frank Blair*, 249–54; Peterson, *Freedom and Franchise*, 170–71; Freehling, *South vs. the South*, 17–32, 36.

Blair's moment of prominent disaster, however, came in 1868, when he was chosen by the Democrats as Horatio Seymour's vice presidential running mate against Grant, whose war-hero status had yet to be marred by the corruption scandals that followed. Blair undoubtedly worsened the inevitable Democratic defeat with his fierce denunciation of congressional Reconstruction, then at the height of its power. His career, however, was rescued with the triumph of his cousin, with whom he had reconciled, as governor. Brown, with the help of Missouri Democrats, secured Blair's election in 1871 as senator, replacing the much-hated symbol of repressive vengeance, Charles Drake. Blair's victory, however, was short-lived. The following year, a stroke crippled him, and he died in 1875.

Brown had made it to the U.S. Senate earlier with his election from the pro-South-purged legislature in 1863. As Radicalism burned itself out, Brown was elected Missouri governor, with Democratic support, in 1870 as a Liberal Republican, promising enfranchisement for freedmen, amnesty for Confederates and their sympathizers, and hostility to Grant administration corruption. The rise of Liberal Republicanism in Missouri turned Brown into a national figure. He was uneasily joined in the movement by Carl Schurz, the German immigrant leader, who had become one of Missouri's senators.

In 1872, the Liberal Republican presidential convention picked Horace Greeley for president and Brown for vice president. The politically enfeebled Democratic party adopted the Liberal Republican ticket as its own, but the result was a disaster, as Grant rode to an easy victory. The Democrat-Liberal Republican ticket was doomed from the start, but scandalous stories about Brown's hard drinking did not help matters. After his defeat, Brown retired to a quiet legal practice and his mathematical studies. He died at his home in Kirkwood, Missouri, in 1885.

James Eads and His Bridge: A Coda to an Era

Throughout his life, James B. Eads was always at the center of the action. Eads was born in Lawrenceburg, Indiana, in 1820 and moved to Missouri as a child.[32] When the booming steamboat business was helping the St. Louis economy explode, steamboats did too, and, in 1842, Eads went into the steamboat salvage

---

32. For Eads, see Florence Dorsey, *Road to the Sea: The Story of James B. Eads and the Mississippi* (New York: Rinehart & Co., 1947); Howard S. Miller and Quinta Scott, *The Eads Bridge*, 2d ed. (St. Louis: Missouri Historical Society Press, 1999); Louis B. How, *James B. Eads* (Boston: Houghton Mifflin, 1900); and Henry Petroski, *Engineers of Dreams: Great Bridge Builders and the Spanning of America* (New York: Alfred A. Knopf, 1995), 22–65.

business. To accomplish this work, he invented a diving bell that allowed his employees, himself included, to stay at the bottom of the river for long periods. Within a decade's time he was rich, but his health was broken. Eads was a man of extraordinary will who drove himself with an almost maniacal intensity, but in a lifelong pattern, he would conclude his efforts with near-complete collapse that caused him to take extended rest, usually doctor-ordered. These "rests" typically took him on extensive tours of Europe, where he relentlessly examined river projects.

During the Civil War, Eads applied his remarkable talents to the Union cause, building in a Carondelet shipyard some of the earliest ironclad boats for the Union navy. These boats proved decisive in Grant's victory at Vicksburg in 1863.

After the war, with no previous experience, Eads turned to bridge-building. St. Louisans had long called for a bridge across the Mississippi, but the Civil War had interrupted this effort. The engineering problems that Eads faced were extraordinary, but they paled beside the machinations of the political and economic foes of the bridge. Eads, however, was almost as indefatigable a politician as he was an engineer. The Eads Bridge, dedicated on July 4, 1874, brought Eads national fame. It is still universally acknowledged as one of the world's bridge-building triumphs.

Eads represented the almost demonic energy of antebellum St. Louis. As much as any of its politicians, he, too, aspired to shape the city and set it on the path to greatness. But in a sense, his bridge was St. Louis's last hurrah. The intensity of St. Louis's antebellum experience had dissipated. The initiative passed to Chicago, which had played no small part in attempting to thwart Eads's bridge-building efforts.[33] The Eads Bridge became a symbol of the redirection of the nation's energies, as railroad companies refused to use it. Eads's company was bankrupt within four years. In 1881 it fell into the hands of New York robber baron Jay Gould.

In 1870 promoter Logan Reavis loudly proclaimed St. Louis "the Future Great City of the World" in a book of that title and tried to get the nation's capitol moved to St. Louis. Yet that same year St. Louisans were reduced to fakery during the U.S. census to ensure that their city would continue to be seen as

---

33. The mid- to late-nineteenth-century St. Louis-Chicago rivalry has been the subject of a great deal of scholarly debate that does not seem finished yet. Some of the more important discussions are Wyatt Winton Belcher, *The Economic Rivalry Between St. Louis and Chicago 1850–1880* (New York: Columbia University Press, 1947); Christopher Schnell, "Chicago versus St. Louis: A Reassessment of the Great Rivalry," *Missouri Historical Review* 71 (April 1977): 245–65; Adler, *Yankee Merchants*, 117–18, 140–77; William Cronon, *Nature's Metropolis: Chicago and the Great West* (New York: W. W. Norton, 1991), esp. 295–309; and Primm, *Lion of the Valley*, esp. 223–26, 272–97. See also Primm's essay in this volume.

larger than Chicago. And no matter what the boosters said, businessmen knew that their upper midwestern markets had been taken over by Chicago, as had the New York City capital that had poured into St. Louis before the Civil War.[34]

In 1875, Missouri adopted a new constitution. This did not mark the end of the repressive measures associated with the Civil War; those measures had fallen before the end of Gratz Brown's governorship. The new constitution did, however, mark the end of an era. It was a conservative document—almost, in its essence, a tax-limitation document, assuring that the state would not have sufficient funds to play a large role in the lives of Missourians. It marked a retreat to localism. The city of St. Louis was divided from St. Louis County and would henceforth go its own way as an independent political entity.[35] It would continue to grow at a remarkable pace for decades to come. There was much money to be made. Smart and energetic businessmen rebuilt their lost northwestern markets by reorienting themselves to a southwestern strategy in Tennessee, Kentucky, Arkansas, and Texas. Immigrants still came. But the frenzied antebellum energy, the sense that life was fluid and that history's wheel was still in spin, gradually declined. St. Louis had much of which to be proud, but heady plans for serving as the center of the new American empire were gone forever.

---

34. Logan Uriah Reavis, *Saint Louis, the Future Great City of the World* (St. Louis: By order of the St. Louis County Court, 1870). For the 1870 census, see Primm, *Lion of the Valley,* 272, and Jeanette C. Lauer and Robert H. Lauer, "St. Louis and the 1880 Census: The Collective Shock of Failure," *Missouri Historical Review* 76 (January 1982): 151–63.

35. Lawrence O. Christensen and Gary R. Kremer, *A History of Missouri* (Columbia: University of Missouri Press, 1997), 4:1–4; Isidor Loeb, "Constitutions and Constitutional Conventions in Missouri," *Missouri Historical Review* 16 (January 1922): 189–246. See also Thomas S. Barclay, *The Movement for Municipal Home Rule in St. Louis* (Columbia: University of Missouri Press, 1943), and Thomas S. Barclay, *The St. Louis Home Rule Charter of 1876* (Columbia: University of Missouri Press, 1962).

African Americans in Henry Shaw's St. Louis

Antonio F. Holland

The lot of African Americans in Henry Shaw's St. Louis was complex. The city's position as a crossroads of transportation and commerce, its French and Spanish beginnings, and its Catholic traditions all made the African American experience here different from that of many other southern border urban centers. Additionally, St. Louis's legacy as a slave city was undermined from the very beginning by the existence of a population of free blacks who established vital religious and cultural institutions that enriched black life. From early colonial settlers living under the rule of the French *Code Noir* to the "colored aristocracy" of the antebellum years, and through the prominent black churchmen and educators of the later nineteenth century, African American St. Louisans established a strong cultural life in the shadow of slavery.[1]

Slavery

The St. Louis to which Henry Shaw came in 1819 was a town still very much influenced by its French and Spanish colonial history. The introduction of Africans into this region had occurred in the early eighteenth century, as France tried to colonize and profit from its vast lands between the Mississippi River and the Rocky Mountains. In 1719 the French explorer Des Ursins, seeking silver mines in Upper Louisiana, brought five enslaved blacks from St. Dominique (later Haiti) as workers. The following year, Phillipe Renault temporarily brought five hundred black slaves from St. Dominique into what is now Missouri to work in the lead mines. Slavery flourished, and Africans soon replaced enslaved Native American inhabitants as the colony's chief forced-labor

---

1. On African American Missouri generally, see Lorenzo J. Greene, Gary R. Kremer, and Antonio F. Holland, *Missouri's Black Heritage,* rev. ed. (Columbia: University of Missouri Press, 1993); portions of this essay have been adapted from the first several chapters of this book.

supply, even after France's 1762 cession of Louisiana to Spain. In 1772, eight years after its founding, St. Louis had a total population of 577, including slaves. Within a year, the village's population had grown to 444 whites and 193 black slaves.[2]

Louisiana's French government had controlled its black slave population under the *Code Noir*, or Black Code. Although black slaves were considered property to be brought and sold, the code did provide explicit sanctions against certain forms of inhumane behavior. As a rule, black slaves in Upper Louisiana were treated better than their counterparts in the Deep South or in Central and Latin America, but they still depended on a sparse diet of cornmeal with supplements of meat and fish, and they lived generally in separate, poorly built housing. During the winter of 1792, according to a Spanish official then in St. Louis, extreme cold led to the deaths of several animals and people. Of the twelve human casualties, most were slaves.[3]

The free black population slowly increased as masters manumitted some for faithful service, others because they were mulatto children whose white parents had bought their freedom, and still others as they purchased their own freedom directly. Some, like Jeanette Forchet of St. Louis, acquired significant amounts of property—which in her case included a house on Second Street and a lot in the Grande Prairie Common Fields.[4] In fact, it was not unusual for free women of color to own property in colonial St. Louis. One, known as Françoise, married well and held property in common with her husband. She later inherited all of her white father's property, including about two hundred acres of land. Esther, the mistress and business partner of St. Louis merchant Jacques Clamorgan, owned an entire city block, two other lots with houses, and two farms north of St. Louis. The names of still others are lost to history. For these women, property ownership represented economic security. The house

---

2. Harrison A. Trexler's *Slavery in Missouri, 1804–1865* (Baltimore: Johns Hopkins Press, 1914) provides an overview of slavery in Missouri, but is marred by racial bias. Lloyd A. Hunter, "Slavery in St. Louis 1804–1860," *Bulletin of the Missouri Historical Society* 30 (July 1974): 233–65, and Arvarh E. Strickland, "Aspects of Slavery in Missouri, 1821," *Missouri Historical Review* 65 (July 1971): 505–26, also provide interesting information on black slavery. For more on Native American slavery and its abolition by Upper Louisiana's Spanish government, see Russell Magnaghi, "The Role of Indian Slavery in Colonial St. Louis," *Bulletin of the Missouri Historical Society* 31 (July 1975): 264–72.

3. William E. Foley, *The Genesis of Missouri: From Wilderness Outpost to Statehood* (Columbia: University of Missouri Press, 1989), 114–15; James Neal Primm, *Lion of the Valley: St. Louis, Missouri, 1764–1980*, 3d ed. (St. Louis: Missouri Historical Society Press, 1998), 23–24. For more on slavery in Spanish colonial Louisiana, see Gilbert C. Din, *Spaniards, Planters, and Slaves: The Spanish Regulation of Slavery in Louisiana, 1763–1803* (College Station: Texas A & M Press, 1999), and Gwendolyn Midlo Hall, *Africans in Colonial Louisiana: The Development of Afro-Creole Culture in the Eighteenth Century* (Baton Rouge: Louisiana State University Press, 1992).

4. Foley, *Genesis of Missouri*, 116–17.

often served as a place of business and a source of income, as well as the prerequisite for acquiring lands in the city's common fields. Under Spanish law a free woman could keep in her own name what property she brought to a marriage, as well as half of what was acquired during the marriage in case of divorce. On the death of the husband, half of his property went to the wife and half to any children. With such protections in place, free women of color constituted an important part of the colonial economy.[5]

While free people of color were more concentrated in St. Louis than elsewhere, they never constituted a large segment of the population in Upper Louisiana. For example, in 1791 there were 23 free mulattoes and 15 free blacks in St. Louis. In 1800, 70 of Upper Louisiana's 77 free persons of color lived in St. Louis. With a total of 1,191 African slaves living in the territory, these free blacks constituted only 6 percent of Upper Louisiana's black population.

Free blacks worked in a variety of occupations. In addition to farming and lead mining, some worked as rowers in the Mississippi River boat trade, while others, like York of Lewis and Clark fame, may have been hunters and trappers. Some freemen possessed skills in trades; one of these was Joseph Barboa, a master mason who worked for the Spanish government on the fortification at St. Louis. A few, like the West Indian–born entrepreneur Jacques Clamorgan, rose to great wealth and power. But most worked in basic positions as domestic servants, dockhands, craftsmen, warehousemen, and mill workers.[6]

The political situation of Upper Louisiana was profoundly changed as a result of the actions of black subjects in another French colony: Haiti. In 1791, inspired by the rhetoric of the American and French revolutions, Haitian slaves rebelled against their white masters. For more than a decade, led by the great black generals Toussaint L'Ouverture, Jean-Jacques Dessalines, and Henri Christophe, they defeated the French, Spanish, and British armies sent to subdue them. Napoleon, who had envisioned an American empire with Haiti as its capital, abandoned his dream in 1803 after having lost forty thousand of his best troops to yellow fever and to battle. When the United States showed interest in the purchase of New Orleans, he offered the entire Louisiana territory.

The U.S. purchase of Louisiana in 1803 was followed by a dramatic increase in the population of that territory. As the number of American settlers moving

---

5. Judith A. Gilbert, "Esther and Her Sisters: Free Women of Color as Property Owners in Colonial St. Louis, 1765–1803," *Gateway Heritage*, 17:1 (Summer 1996): 14–23. See also Katharine T. Corbett, *In Her Place: A Guide to St. Louis Women's History* (St. Louis: Missouri Historical Society Press, 1999), 3–19.

6. Primm, *Lion of the Valley*, 58. We know more about labor in Lower Louisiana in the same period; see Kimberly S. Hanger, *Bounded Lives, Bounded Places: Free Black Society in Colonial New Orleans* (Chapel Hill: Duke University Press, 1997).

"Inventory and appraisal of property belonging to Jeanette Forchet, a free Negress" (May 30, 1790; photograph by David Schultz). Like other free women of color in colonial St. Louis, Forchet enjoyed some legal protections that would not have been hers under the subsequent American regime. This inventory documents possessions held in her name at the time of the death of her second husband. *Missouri Historical Society*

into new land increased, so did the number of slaves they brought with them to conquer the frontier. At the time of the Louisiana Purchase there were 10,340 free persons living in Missouri, in addition to 1,320 slaves. By 1810, Missouri's free population had grown to 20,845, an increase of more than 100 percent. During that same period the black population, growing at a much faster rate than that of the general population, rose to 3,618. The majority of Missouri slaves were located in the Mississippi and Missouri river valleys, and most of those lived in St. Louis. The commander of the Missouri territory, Captain Amos Stoddard, reported 667 slaves living in the district of St. Louis in 1804—some 25 percent of the district's entire population. That number increased to 740 by 1810.[7]

Expecting to gain their freedom like those blacks living in the Northwest Territory, black slaves in Missouri grew restless. Local slaveowners like Auguste Chouteau warned Stoddard that blacks and troublesome whites were consorting, and recommended a set of regulations to control the slaves. Whites' fears of insurrection echoed those of their colonial precursors. The French and Spanish black codes had made it illegal for slaves to leave an owner's property without his permission, to carry guns, or to strike their masters, and the Spanish Lieutenant Governor Francisco Cruzat had found it necessary in 1781 to issue new ordinances to prevent the numerous unauthorized night meetings of slaves. The slaveowners of the new Missouri Territory had even more reason to be fearful. The rebellion of L'Ouverture and his followers in Haiti had struck fear in the hearts of southern slaveowners. In Virginia in 1800, Gabriel Prosser had planned an (unsuccessful) effort to lead a band of over one thousand slaves on Richmond. Many Missourians had migrated from Virginia; they carried with them memories of Prosser—and of the code that Virginians had enacted to forestall another revolt. One of the first acts of Missouri's new territorial government was a new series of black or slave codes, patterned closely after those of Virginia, which were considered the harshest in the United States. These codes, which applied to "any person who shall have one-fourth part or more of negro blood," made no distinction between slaves and other property. Slaves could not testify in court against whites. As in the colonial period, they could not leave their owner's property without his permission or own or carry guns.[8]

Between 1808 and 1818, St. Louis officials passed a series of additional laws that prevented slaves from drinking, holding public meetings, or mixing at

---

7. Hunter, "Slavery in St. Louis," 235–36.
8. Primm, *Lion of the Valley*, 72; Foley, *Genesis of Missouri*, 116, 254. See also William E. Foley, "Slave Freedom Suits before Dred Scott: The Case of Marie Jean Scypion's Descendants," *Missouri Historical Review* 79 (October 1984): 4; Trexler, *Slavery in Missouri*, 58; *Territorial Laws*, vol. 1, chap. 3.

night with either free blacks or whites. By the statute of 1818 any slave found out of doors after 9 P.M. could be whipped or put in jail. White St. Louisans probably felt that they had just cause for increasing their control over the black population. In 1805 Pierre Chouteau believed that a slave had started the fire that burned down his home in St. Louis. In 1818 a St. Louis slave, Elijah, was charged with trying to poison his master's family.[9]

By 1820, the year after Henry Shaw's arrival in St. Louis, the Missouri Territory claimed an unofficial total of some 66,000 residents, of whom 10,000 were slaves; the free black population, meanwhile, had dropped to 375 from 600 in 1810. St. Louis County claimed 1,860 slaves (18 percent of its population) and only 200 free blacks.[10] In 1819, in a rehearsal of the great crisis that would eventually split the country into civil war, New York Representative James Tallmadge fought unsuccessfully to exclude the further introduction of slavery into the new territory, and to free, when they reached the age of twenty-five, those children of slaves born in Missouri after it attained statehood. The Missouri Compromise of 1820, which at last admitted Missouri as a slave state, temporarily averted crisis but set the stage for the dramatic events of the 1850s and 1860s.[11]

Slavery was, above all else, an economic institution. Slave masters were interested in getting as much work out of their slaves as they could for as long as possible. The life of the slave was determined largely by this fact. Although the majority of Missouri slaves worked as field hands on farms, as a group they possessed a wider range of skills and occupations than their counterparts in the Deep South. Many—especially in St. Louis—were employed as valets, butlers, handymen, carpenters, common laborers, maids, nurses, and cooks. Missouri and St. Louis masters often hired their slaves out when they were not using them, earning cash payment for their slaves' labor. In addition, this system transferred the cost of room and board to the person hiring the slave. When railroads began moving through Missouri, black slaves were called upon to work at the rate of twenty dollars a month; brickyards paid a similar sum. Many slaves were hired out to the owners of riverboats, who worked them as deck hands, cabin boys, or stevedores for approximately fifteen dollars a month—lower wages than those paid whites for similar work.

For the slave, of course, no such economic calculus could overcome the essential inhumanity at the heart of the system. Supposedly protected under the law from excessively harsh treatment, slaves were in practice unsuccessful in

---

9. Foley, *Genesis of Missouri*, 254–56.
10. Ibid., 253–54; Primm, *Lion of the Valley*, 115; Hunter, "Slavery in St. Louis," 236.
11. Richard C. Wade, *Slavery in the Cities: The South, 1820–1860* (New York: Oxford University Press, 1964), 239.

challenging the private right of owners to punish their property. The spectacle of female slaves forced to wear iron chains while doing kitchen work was not uncommon in St. Louis. In 1839, a St. Louis master left a female slave hanging by her thumbs.[12]

The system of slavery was able to exist only by forcing blacks to stand in fear of whites. Because slave masters, as a group, believed that a fearful slave was also an obedient slave, they willingly used whatever means they could to instill fear. On the other hand, slave masters also realized that their ill treatment of slaves would inspire black resistance. Consequently, they sought an elaborate system of laws to protect themselves, their families, and their communities against any semblance of black rebellion. They tried to enlist the support of non-slaveholding whites. Through religion, philosophy, law, and social practice, they tried to convince the poorest white person that he was better than the most cultured and intelligent black. Whites of all classes were encouraged to believe that they could someday enhance their social status by becoming slaveowners, even though seven out of eight Missouri families never owned slaves.

Missouri's slave laws illustrate the general attitude toward blacks during the antebellum period. As personal property, slaves were taxed, bought, and sold. As to their needs—food, clothing, shelter, health care—in most cases these items rested in the hands of the owners. Slave marriage was forbidden. Sometimes a man and woman "took up" with each other, while at other times they were ordered to live together by the master. As far as whites were concerned, the sale of one of the partners ended the relationship; slave "marriages" had no legal standing.

Nor were these the only statutes directed against blacks. Slaves found guilty of crimes such as conspiracy, rebellion, or murder were to be put to death. As to lesser infractions, slaveowners enjoyed a great deal of discretion. Often a slave was whipped in the presence of other slaves so that the master could provide a dramatic lesson to disobedient slaves. Such punishment could cause blisters, draw blood, and leave permanent scars. Usually, however, the master tried to avoid permanently marking a slave, so as to protect his economic investment. A scarred slave would be regarded by a prospective buyer as a problem, and his potential value would be lessened accordingly.[13]

State laws further forbade sexual relations between slave men and white women, and black or mulatto men who assaulted or attempted to assault a white woman were subject to death or mutilation, often by castration. Rape of a slave

---

12. Primm, *Lion of the Valley*, 180; Hunter, "Slavery in St. Louis," 248–49; Wade, *Slavery in the Cities*, 192–93; *St. Louis Missouri Republican*, May 24, 1824.
13. Hunter, "Slavery in St. Louis," 248.

woman by a white man was not considered an offense against the woman or a violation of the law but a case of trespassing on the master's "property." Bondsmen guilty of striking a white person were to be punished, at the discretion of the justice of the peace, with a beating not to exceed thirty-nine lashes. Thirty-nine lashes was also the penalty for slaves who disturbed church service by "noise, riotous or disorderly conduct." Twenty-five lashes and a jail term awaited any person providing liquor to a slave.[14]

By 1847, even as the numbers of slaves within St. Louis dwindled dramatically, Missouri slaveowners sought still greater powers to control unrest. Mindful of the uprisings led elsewhere by Nat Turner, Denmark Vesey, and others, and convinced that such rebellions were fanned by abolitionist literature, slaveowners sponsored an act specifically prohibiting the education of blacks. Under its terms, anyone operating a school for Negroes or mulattoes or teaching reading or writing to any Negro or mulatto in Missouri could be punished with a fine of not less than five hundred dollars and sentenced up to six months in jail.[15]

One of the most important institutions established to ensure the safety of the slaveowners and the community was the system of the slave patrols, groups of community men authorized to monitor the movement of slaves. St. Louis established such patrols in 1811 and 1818. The State of Missouri authorized each county to set up its own patrol in 1823. This legislation was designed to ensure that slaves were not traveling abroad at night without their master's consent. Patrols also visited the slave quarters to guarantee that there were no unlawful assemblies of slaves. Thirty-nine lashes was, again, the punishment for such an illegal meeting.[16]

As the intensity of the antislavery struggle increased in the years immediately preceding the Civil War, slave-control laws grew more numerous still. In most Missouri cities all blacks without a pass had to be off the street by 9 P.M. unless on business for their master. Special permission had to be secured from the master for all special meetings, and all such persons had to be home by 10 P.M. Passes were good for twenty-four hours only, and the city constable had to see that meetings were orderly.[17]

The almost insatiable demand for black slaves in the South meant that there was always a buyer from that region ready to purchase a Missouri slave. Several slave dealers carried on the trade in Missouri, buying and selling blacks for use

14. Ibid., 257–58.
15. Ibid.; Donnie D. Bellamy, "The Education of Blacks in Missouri Prior to 1861," *Journal of Negro History* 59 (April 1974): 143–57.
16. Trexler, *Slavery in Missouri*, 183.
17. Hunter, "Slavery in St. Louis," 256–58.

in southern cotton, rice, sugar, and indigo fields. St. Louis was the largest slave mart in the state. Slave dealers there advertised in various counties for slaves. St. Louis branches of southern-based slave-trading concerns such as Blakey and McAfee kept an agent in the state capital in the early 1850s. Offering "highest prices for Negroes of every description," they boasted of their facilities as being "well suited for the boarding and safe keeping of Negroes sent to this market for sale." Competing with Blakey and McAfee for the Missouri trade was yet another southern company, Bolton, Dickens and Company. In some cases, dealers sold slaves on commission, boarding them until sold at owner's risk and expense. A typical newspaper advertisement posted for sale "a good negro man age 52."[18]

No slaveholding state admitted to the business of breeding slaves for sale; the question of Missouri's involvement in that practice remains debatable. Despite this, William Wells Brown, an escaped slave who gained international renown as an antislavery speaker and novelist, argued that "Missouri . . . is very much engaged in raising slaves to supply the southern markets." Like Brown, hundreds of other slaves (including several belonging to Henry Shaw) deprived their masters of their investment in human property by running away, often carrying with them some of the master's property. Local newspapers typically featured notices and descriptions of the runaways.[19] Many slave masters felt that slave escapes were being facilitated by the Underground Railroad, a system by which free blacks and sympathetic whites provided assistance for runaways fleeing north. Although it seems likely that the Underground Railroad did help some Missouri slaves to escape, it is unlikely that it had the effect on Missouri slavery that contemporary masters believed.

Free Blacks

Free blacks lived beside their enslaved brethren throughout Missouri's slave period. In 1860, 3,572 free blacks resided in the state, half of them in St. Louis. The question of "free licenses"—permits needed by free blacks in order to reside in a Missouri county— keenly reflects the attitude of policymakers toward blacks in general and free blacks in particular. Blacks were presumed to be slaves until they could offer proof to the contrary. A freedman brought before a justice of the peace who was unable to persuade the court that he was free could

---

18. Ibid., 260–62; *St. Louis Missouri Republican,* January 7, 1854.
19. William Wells Brown, *From Fugitive Slave to Free Man: The Autobiographies of William Wells Brown,* ed. William L. Andrews (New York: Mentor Books, 1993), 81; Trexler, *Slavery in Missouri,* 50–51.

be jailed as a runaway and sold into slavery. In 1846 the constitutionality of the pernicious free-license law was sustained by Judge John M. Krum of the St. Louis Circuit Court, who denied the claim that blacks were citizens. In 1857 the same reasoning would, of course, be used by the United States Supreme Court in rejecting the St. Louisan Dred Scott's insistence upon equal protection under the laws of the United States. The decision deprived free blacks of any illusions they may have had as to citizenship and civil rights under the Constitution.[20]

When legal means seemed insufficient to restrict free blacks' behavior, whites sometimes turned to extralegal means. In April 1836, a St. Louis mob lynched Francis McIntosh, a free mulatto steamboat cook from Pittsburgh who had allegedly stabbed one constable to death and seriously wounded another. The mob lashed McIntosh to a tree and burned him alive on the outskirts of the city. At the inquest into McIntosh's murder, presiding judge Luke Lawless condoned the self-appointed executioners' action as "beyond the reach of human law," and blamed abolitionists such as Elijah Lovejoy for causing the crime. The brutal lynching and the inaction of authorities to bring the lynchers to justice gave St. Louis a national reputation for lawlessness.

Five years later, civic-minded St. Louisans saw an opportunity to redeem their city's notorious reputation. In May 1841, St. Louis authorities arrested three free blacks—Charles Brown, James Seward, and Alfred Warrick—and a slave, Madison Henderson, charging them with bank robbery, arson, and the murders of two white bank tellers. In order to secure a fair hearing for the accused, the court appointed three prominent white attorneys to serve as their defense. The four accused men, however, were found guilty and sentenced to death by hanging. In July 1841 an estimated crowd of twenty thousand—approximately three-quarters of the St. Louis population—watched the execution of the four at nearby Duncan Island. Excursion steamboats carried passengers the short distance out to the island to witness the event.[21]

Despite these legal restrictions and other obstacles, the free black class increased its numbers in a variety of ways. One, of course, was escape. Generally, however, a slave gained his freedom only with the consent of his master. Two motives generally entered into the act of liberating a slave: financial consideration and sentiment. Slaves seeking to purchase their freedom had to negotiate

20. Don E. Fehrenbacher, *The Dred Scott Case: Its Significance in American Law and Politics* (New York: Oxford University Press, 1978), 249–53. Originally, Dred Scott and his wife, Harriett, suited separately for their freedom, but the Missouri Circuit Court at St. Louis combined the suits. See also Kenneth C. Kauffman, *Dred Scott's Advocate: A Biography of Roswell M. Field* (Columbia: University of Missouri Press, 1996).

21. Janet S. Herman, "The McIntosh Affair," *Bulletin of the Missouri Historical Society,* 26 (January 1970): 123–43; Mary E. Seematter, "Trials and Confessions: Race and Justice in Antebellum St. Louis," *Gateway Heritage* 12 (Fall 1991): 36–47.

African Americans in Henry Shaw's St. Louis 61

Dred Scott (oil on canvas by Louis Schultze, 1888). Scott's unsuccessful effort to gain freedom, based on his earlier residence in free territory, expanded from a local to a national story during its long path through the American judicial system. Years after his death, the Missouri Historical Society commissioned this posthumous portrait. *Missouri Historical Society*

a purchase price higher than the value of their continued servitude. The cost of freedom depended upon several variables, including sex, age, health, skills, and the availability of a ready market for slaves. Against great odds, a number of slaves did manage to pay such a cost. One St. Louis slave, Jesse Hubbard, accompanied his master to California and returned with fifteen thousand dollars. Subsequently, Hubbard purchased his freedom and bought a farm in St. Louis County.[22]

In rare instances, Missouri's free blacks would themselves become slaveholders—often, in order to allow their slaves an opportunity to earn money for self-purchase. One of the most famous of such cases was that of John Berry Meachum. Born a slave in Virginia, Meachum worked under a skilled craftsman from whom he learned carpentry, cabinet making, and coopering. He earned enough money to purchase his freedom. When his enslaved wife's master moved to St. Louis, he pulled up stakes for Missouri. There he found employment and was soon able to purchase his wife and children. Meachum's industriousness allowed him to save enough money to open a barrel factory. Between 1826 and 1836, he purchased approximately twenty slaves whom he employed in his factory until they had learned a trade and saved enough money to buy their freedom.

In addition to being a businessman, Meachum was also an ordained minister. He was the founder of the First African Baptist Church of St. Louis, located on Third and Almond streets. In the years before the 1847 legislation outlawing education for blacks, Meachum was in the forefront of efforts to educate blacks, both slave and free. Initially he taught slaves and free blacks to read and write under the guise of conducting a Sunday school. When whites became aware of what he was doing, Meachum switched tactics. He built a steamboat, equipped it with a library, and anchored it in the middle of the Mississippi River—a location subject to federal but not state law. Each morning students were transported to his school by means of a skiff; once on board they learned reading, writing, and arithmetic, in defiance—but outside the reach—of state law. Meachum's "Freedom School" continued until his death in the late 1850s.

There were other efforts to educate blacks in antebellum Missouri. Timothy Flint, a northern white missionary minister, conducted a school in St. Charles from 1816 to 1826. A number of schools for blacks run by St. Louis Catholics operated throughout the antebellum period. Hiram K. Revels, later elected the nation's first African American senator in Reconstruction-era Mississippi (in a seat previously held by Jefferson Davis), opened a school in St. Louis in 1856, enrolling approximately 150 free blacks and slaves at a cost of one dollar per

---

22. Trexler, *Slavery in Missouri*, 219–20.

pupil per month.[23] Revels did not have to resort to the extremes of Meachum, because the control of the machinery of slavery had greatly weakened by the late 1850s.

One of the most important institutions in the Missouri free black community was the church, which provided freedmen with a source of stability and social strength in an otherwise hostile society. The black churches provided one of the few forums in which potential black leaders could develop and refine their leadership skills. Unfortunately, the same sentiment that led blacks to establish their own churches caused whites to fear them. Often blacks who wished to worship had to do so in white churches, physically separated from the rest of the congregation. St. Louis, home to nearly half of the state's free blacks, was the only place in the state where the black church achieved any measure of success. By 1860 the city boasted three Baptist and two Methodist churches founded by blacks, in spite of an 1847 state law requiring a county official to attend all religious services conducted by blacks.

Employment was also a problem for free blacks. Some stayed with their masters after receiving their freedom and continued as farm workers or common laborers. Skilled freedmen stood a better chance of finding work, but even they found employment difficult in rural areas. Because of these restrictions, free blacks in the state tended to migrate to its cities, especially St. Louis. In 1860, the Gateway City's 1,755 "free men of color" represented nearly half of the state's total free black population. These former slaves worked as barbers, wagoners, blacksmiths, carpenters, house servants, cooks, waiters, draymen, stonemasons, watchmen, carriage drivers, painters, gardeners, hostlers, stable keepers, store owners, chambermaids, washerwomen, ironers, and seamstresses. Undoubtedly, Missouri's most famous black seamstress was Elizabeth Keckley, who purchased her and her son's freedom in 1855, eventually leaving the city to become an employee and confidante of First Lady Mary Todd Lincoln. Pursuing a less urbane, if more glamorous, occupation was the famous black mountain man Jim Beckwourth, who in 1850 discovered an important pass through the Sierra Nevada Mountains.[24]

---

23. Judy Day and M. James Kedro, "Free Blacks in St. Louis: Antebellum Conditions, Emancipation, and the Postwar Era," *Bulletin of the Missouri Historical Society* 30 (January 1974): 117–35; N. Webster Moore, "John Berry Meachum (1789–1854): St. Louis Pioneer, Black Abolitionist, Educator, and Preacher," *Bulletin of the Missouri Historical Society* 29 (January 1973): 96–103; Donnie D. Bellamy, "Free Blacks in Antebellum Missouri, 1820–1860," *Missouri Historical Review* 67 (January 1973): 224.

24. Elizabeth Keckley, *Behind the Scenes, or Thirty Years a Slave and Four Years in the White House* (1868; reprint, New York: Oxford University Press, 1988), 63; *The Life and Adventures of James P. Beckwourth as Told to Thomas D. Bonner,* ed. Delmont R. Oswald (Lincoln: University of Nebraska Press, 1972).

Elizabeth Keckley (wood engraving; frontispiece of Keckley's autobiography, *Behind the Scenes, or Thirty Years a Slave and Four Years in the White House* [New York: G. W. Carleton & Co., 1868]). Keckley, a slave in St. Louis, purchased her freedom and then parlayed her skills as a seamstress into a career that included service to First Lady Mary Todd Lincoln. *Missouri Historical Society*

Nowhere in Missouri could one find more of these individual success stories than in St. Louis. As local free blacks found economic opportunity, and as others migrated to the city in search of the same, there emerged a class of families, many of them mulatto, that Cyprian Clamorgan termed "the Colored Aristocracy." Clamorgan, himself a freedman, claimed in 1858 that free blacks in St. Louis controlled several millions of dollars in real and personal property. According to Clamorgan, Mrs. Pelagie Nash owned nearly the whole block in which she lived. Mrs. Sarah Hazlett, a widow, possessed a significant fortune. Samuel Mordecai, with a business at Fourth and Pine streets, had amassed thousands of dollars. Albert White, a barber who came to St. Louis with $15,000, took his wife to California and returned with an even bigger fortune. William Johnson, a realtor who started with a barbershop in 1840, owned a substantial estate by 1858. A cattle dealer, Louis Charleville, expanded into business and real estate investments; James Thomas, a barber, wrote a memoir detailing his own rise to wealth. The homes of the colored aristocracy were, according to Clamorgan, clustered around several streets: Seventh, Rutger, Third between Hazel and Lombard, Fourth and Pine, and Fifteenth near Clark Avenue. Some looked down upon their darker brothers and sisters, tending to marry other mulattoes.[25]

The fact that a number of free blacks were able to rise above the obstacles placed on them by society only heightened the hatred and fear that many Missouri slave masters held for free blacks. The clearest indications of this hostility were the continuous attempts to colonize freedmen outside of the state and even the nation. The American Colonization Society, established by slaveholders in 1816, founded Liberia to rid the country of free blacks. Many Missouri emancipationists were willing to endorse the abolition of slavery in the state only if the freed blacks were transported out of the country. Emancipationists were particularly eager to send Missouri freedmen to Liberia. Most free blacks, however, claimed America as their home and resisted emigration; the colonization movement in Missouri, as elsewhere, proved a failure.[26]

Whether wealthy and successful as were the St. Louis "colored aristocrats," or struggling as were so many other free blacks, in the long run freedmen prospered in Missouri, offering a living refutation of their alleged inferiority and inability to live in a predominantly white society. By building churches and

---

25. Cyprian Clamorgan, *The Colored Aristocracy of St. Louis*, ed. Julie Winch (Columbia: University of Missouri Press, 1999), 73–79, 86–89, 95–98. See also James Thomas, *From Tennessee Slave to St. Louis Entrepreneur: The Autobiography of James Thomas*, ed. Loren Schweninger (Columbia: University of Missouri Press, 1984).

26. Bellamy, "Persistence of Colonization in Missouri."

schools, entering into business, and putting to use the skills learned in slavery, they moved toward first-class citizenship.

Civil War

For blacks, slave or free, the Civil War was the true American Revolution. Despite a variety of impediments to black recruitment, more than 180,000 blacks, or about 10 percent of total enlistments, served in the Union cause. These men are credited with taking part in five hundred military actions and nearly forty major battles. More than 20 percent—almost 37,000—of their number gave their lives for the cause. Another 29,000 blacks served in the Union navy, representing 25 percent of that service's manpower. Seventeen black soldiers and four black sailors were awarded the Congressional Medal of Honor, the nation's highest military decoration.

The army began recruiting the first black regiment in Missouri in June 1863 at Schofield Barracks in St. Louis. On June 10, the *St. Louis Tri-Weekly Democrat* announced that more than three hundred blacks had enlisted. By the end of the year the ranks of the First Regiment of Missouri Colored Infantry—later the Sixty-second U.S. Regiment of Colored Infantry—were virtually complete. By January 1864 another regiment was being organized. According to General John McAllister Schofield, commander of the Department of Missouri, the First Missouri Infantry of African Descent and the Second Missouri Infantry of African Descent (later designated the Sixty-fifth U.S. Regiment of Colored Infantry) were among the 4,486 officers and men stationed at Benton Barracks, in north St. Louis. By February 2, 1864, forty-six assistant provost marshals in as many towns had enlisted 3,700 blacks. The City of St. Louis provided 670 recruits.[27]

During the war thousands of slaves ran away. In 1860 there had been more than 100,000 slaves in Missouri. By 1862 the slave population had dropped to 85,000 and by 1864 there were only 22,000. In 1860 a young male slave could cost thirteen hundred dollars; by 1864 the same slave would bring only one hundred dollars. As the end of the war drew near, many black and white leaders were aware that slavery was a dying institution and tried to arrange educational opportunities for slaves and free blacks alike. Black and white educational efforts

---

27. John W. Blassingame, "The Recruitment of Negro Troops in Missouri during the Civil War," *Missouri Historical Review* 68 (April 1964): 326–38; Michael Fellman, "Emancipation in Missouri," *Missouri Historical Review* 83 (October 1988): 36–56. The black soldiers and white officers of the Sixty-second and Sixty-fifth U.S. Regiments of Colored Infantries gave the moneys to establish what is now Lincoln University of Missouri.

on behalf of blacks were so large in St. Louis that a black board of education was established. The unofficial board had charge of four schools with a total of four hundred students. By 1865 the system had eight teachers and six hundred pupils.

In 1865 the Western Sanitary Commission, a white benevolent association, operated a high school in St. Louis for about fifty blacks in the basement of a local church and aided other black schools with a total enrollment of fifteen hundred. The commission also organized classes, mostly in reading and writing, for black soldiers at Benton Barracks. The officers of the black outfits, many of them college-trained, often taught the black soldiers around the campfires.

The federally sponsored Freedmen's Bureau was also effective in offering financial support for local black education near the end of the war. During the war years, however, the most important single force supporting black education was probably the American Missionary Association. In the late 1850s, the AMA had unsuccessfully tried to convert the Missouri slaveholder to the abolitionist position. Forcefully and violently driven out of Missouri at the start of the Civil War, the AMA returned in 1862. This time the main thrust of the organization was to provide the former slaves with a Christian education. Despite black eagerness to receive education, the AMA encountered constant opposition to its efforts. In 1863 an AMA school that served sixty black pupils in St. Louis was burned by a group of white youths. The school had been open for only three days. In a new building, the AMA school averaged more than one hundred pupils daily. The AMA also operated a Sunday school with eight hundred black pupils. Still, only three hundred fifty pupils—some 10 percent of the eligible youth—attended daily classes. In St. Louis and other localities, blacks raised money for teachers' board, paid rent on school buildings, and, insofar as they were able, paid teachers' salaries.[28]

Reconstruction (1865–1877)

Throughout the former slave states, blacks believed a new day had dawned. The era of the war and of slavery had turned their world upside down. Without land, money, or education, blacks moved toward an uncertain future. On January 11, 1865, Missouri slaves were freed, eleven months before the ratification of the Thirteenth Amendment ended slavery nationally.

---

28. Joe M. Richardson, "The American Missionary Association and Black Education in Civil War Missouri," *Missouri Historical Review* 69 (July 1975): 433–48.

Many blacks reacted to white hostility and oppression in Missouri by fleeing the state as soon as they gained their freedom. In fact, there were fewer blacks in the state in 1870 than there had been ten years before. Some of those who stayed tried to find jobs with their former masters or other whites. Often their pay was only room and board.

Even if blacks were able to find employment, they faced other burdens. Although Missouri escaped much of the segregation legislation that would later dominate the South, there were informal codes of behavior designed to ensure that blacks knew their place. For example, streetcar companies in St. Louis prohibited blacks from riding inside their cars. In 1867, Caroline Williams, a black woman, pregnant and holding a baby, tried to board a car only to be shoved into the street by the conductor. She and her husband sued the company. They won their suit but were awarded damages of only one cent. Nevertheless, a principle of law had been established that effectively ended the practice of keeping blacks out of streetcars. Despite the victory, lawless violence could strike blacks at any time; in 1867, a drunken band of whites fired pistols into a black Christmas Eve congregation at St. Paul's AME Church in Columbia. They killed one person and wounded another.

Federal legislation provided a key framework for defining the rights of American blacks. Congressional Radicals at the national level pushed through important amendments to the Constitution: the Thirteenth, ratified in 1865, which abolished slavery forever; and the Fourteenth (1868), which guaranteed blacks equal protection of the laws and all civil liberties enjoyed by white persons. The federal government also assisted the freedmen in more specific ways. One was the establishment in 1865 of the Freedmen's Bank, which encouraged them to save their money in order to secure a stronger economic position in postwar society. In 1868 a St. Louis branch was established under the presidency of the Reverend William P. Brooks, who served for the entire six years of the bank's life. Under Brooks's stewardship, the society grew. Still, while some of the city's former free-black aristocracy had money, most of the recent freedmen had very little. The bank fell victim to the financial crisis that swept the country during the Panic of 1873; it closed its doors in 1874. Another key institution, the Western Sanitary Commission, remained especially active in the area of education; the commission also set up a Freedmen's Orphan Home for abandoned children of slaves.[29]

---

29. For Brooks, see John A. Wright, *Discovering African American St. Louis: A Guide to Historic Sites* (St. Louis: Missouri Historical Society Press and St. Louis Public Library, 1994), 13; for the orphan home, see *Western Sanitary Commission Final Report*, no. 5 (St. Louis: R. P. Studley & Co., 1866).

Politically, black Missourians failed to receive what they desired. When the convention that crafted the new state constitution soundly rejected the inclusion of blacks' right to vote and hold office, blacks responded by organizing Missouri's first black political activist movement: the Missouri Equal Rights League. The league's initial meeting was held at a church on the corner of Green and Eighth streets in St. Louis in October 1865. It was dominated by black St. Louis religious leaders, although freedmen from other parts of the state were present.

The blacks gathered in St. Louis called attention to their plight as freedmen who lacked the rights and privileges of the elective franchise and charged that such a condition was little better than the oppression they had suffered under slave masters. They pointed out that they too had borne arms in defense of the Union and stated that their future safety and prosperity would be best ensured by the state declaring all people, regardless of color, equal before the law.

Before adjourning, League members chose a seven-member executive committee to work toward the achievement of this goal. The committee was charged with the responsibility of providing for a series of mass meetings throughout the state, procuring black speakers, and preparing an address on the plight of black people to the citizens of Missouri.[30]

The Reverend Moses Dickson, a prominent black St. Louisan, was probably the best known of the committee's members. His involvement with the Equal Rights League represented the continuation of a struggle for black freedom that he had begun long before. After traveling through the South from 1840 to 1843 and observing the plight of slaves, Dickson recruited twelve men and formed a group called the Knights of Liberty, which aimed to enlist and arm southern slaves for a general rebellion. At one point, the Knights claimed to have forty-seven thousand members. The group set the year 1856 as the time for revolution. As the year approached and the antislavery struggle increased in intensity, Dickson counseled his followers to wait for what he felt certain would be a civil war. When the war came in April 1861 and black troops were authorized, Dickson and many of his followers took up arms for the Union cause. Later, a fraternal organization known as the International Order of Twelve Knights and Daughters of Tabor was founded to continue the struggle for equality. This organization still exists today.[31]

---

30. Gary R. Kremer, *James Milton Turner and the Promise of America: The Public Life of a Post–Civil War Black Leader* (Columbia: University of Missouri Press, 1991), 19–20. The material on the Missouri Equal Rights League is taken from Professor Kremer's work.

31. Moses Dickson, *Manual of the International Order of Twelve Knights and Daughters of Tabor* (St. Louis: A. R. Fleming, 1891), 1–8. Dickson is the only known source for this exciting story; it is difficult to assess the accuracy of his claims.

Two weeks after the Equal Rights League executive committee was appointed, a twenty-seven-hundred-word *Address to the Friends of Equal Rights* appeared in local newspapers and as a pamphlet. The address was an elaborate expression of the concerns and aspirations that had been voiced at the October organizational meeting. Its major plea was for the right to vote. The petitioners reminded readers that they were citizens of the state and nation and that their work had benefited both. They also recalled that nine thousand black soldiers had "bared their breasts to the remorseless storm of treason, and by hundreds went down to death in the conflict while the franchised rebel . . . the . . . bitterest enemy of our right to suffrage, remained . . . at home, safe and fattened on the fruits of our sacrifice, toil and blood."[32] The address advised that all should take seriously the plea for the right to vote, emphasizing that the fate of black people was the greatest issue before the nation.

The committee hired John M. Langston of Ohio, a well-known black orator and lawyer, to tour the state in support of its petition. Langston began his journey with a talk in St. Louis in November 1865. Langston was not the only black spokesman touring the state. Twenty-six-year-old James Milton Turner became one of the most important black leaders in post–Civil War Missouri. Born a slave in 1839, he gained his freedom in 1843. Educated in St. Louis schools and at Oberlin College in Ohio, Turner was an exciting orator who emerged as secretary of the Missouri Equal Rights League in 1865. Throughout that winter he traveled around the state, especially the southeastern part, advocating education and the ballot for blacks.

The executive committee circulated a petition throughout the state, imploring the legislature to provide suitable schools for black children. It also pushed for an amendment to the constitution that would remove the word "white," thereby guaranteeing the legal equality of all the state's citizens. The petition was signed by four thousand blacks and whites. It was then turned over to the Honorable Enos Clarke, state representative from St. Louis, to present to the legislature. Despite Clarke's strong endorsement of the petition, the legislature refused to act favorably upon it. In fact, the result of all the activism in Missouri was disappointing. Success came only in 1870 when the Fifteenth Amendment to the national Constitution guaranteed the right to vote without regard to "race, color, or previous condition of servitude."

The black citizens who made up the Equal Rights League had known that the fight to secure full political and civil rights was not going to be easy. Accordingly, they placed great emphasis on education to advance the progress of blacks and to refute the arguments of their white opponents against giving blacks the vote.

32. Quoted in Kremer, *James Milton Turner and the Promise of America*, 21.

They realized that many freedmen were unprepared for participatory democracy. The executive committee in its address emphasized, "We mean to make our freedom practical," adding that it saw education as the chief means by which that could be done. Convinced that the responsibilities of citizenship could be best fulfilled by an educated citizenry, the league sought to establish schools for the freedmen wherever possible.

Turner, the Equal Rights League secretary, was the most active and effective black advocate of education for the former slaves. He helped win state support for the Lincoln Institute in 1870, and was a leader in the movement to establish schools for blacks across the state. Nationally, the Radicals still held some sway, and President Grant appointed Turner to the Liberian ministership. Turner held that position from 1871 to 1878, making him only the second black person in the history of the country to become a diplomat.[33]

End of Reconstruction to 1900

Rutherford B. Hayes won the 1876 presidential contest only after agreeing to certain concessions to the South, among which was the withdrawal of federal troops from that region. With reconstruction officially ended, black rights could be violated without fear of federal intervention; the abandonment of the protection of black people's civil rights was complete. Many southern blacks responded to this turn of events by fleeing the South for the "promised land" of Kansas and other points west. St. Louis, located on the Mississippi River, became a way station for the journey west. By early March 1879, the first of the more than six thousand blacks who would come to St. Louis during the next four months arrived. The problems faced by these "exodusters," as the migrants came to be known, were legion. The first obstacle to overcome was to secure money for boat fare up the Mississippi. It cost from three to four dollars per adult to travel from the vicinity of Vicksburg, Mississippi, to St. Louis. Children under ten years of age were transported for half price, and a small amount of baggage was carried free of charge. Consequently, a family of five needed from ten to fifteen dollars just to get up the river—an amount that, in many cases, blacks had to raise by a hasty and unprofitable sale of most of their household goods.

They arrived in St. Louis with their money spent and no way to secure passage to Kansas. The black community of St. Louis quickly responded to the exodusters' needs. The prominent black leader Charlton H. Tandy organized and

33. Kremer, *James Milton Turner and the Promise of America*, 18–26.

spearheaded the relief effort. Tandy helped the men find jobs and arranged for impromptu food and shelter for the first group of several hundred who arrived in St. Louis on March 11, 1879. Tandy's efforts to solicit the aid of whites in St. Louis were largely unsuccessful. Indeed, Mayor Henry Overstolz discouraged attempts to aid the migrants, lest other destitute southern blacks be attracted to his city. Realizing that whites were going to be of little help, Tandy organized a group of fifteen persons, later enlarged to a committee of twenty-five, called the Colored Relief Board. The committee assumed responsibility for seeing to the needs of the migrants, including the arrangement and supervision of their transportation to Kansas.

Most of the money raised in this relief venture came through the black churches of St. Louis. Between March 17 and April 22 alone, St. Louis blacks provided the exodusters with nearly three thousand dollars' worth of goods and services, making it possible for the vast majority of them to move on to Kansas. Besides Tandy, the Reverend John Turner and Reverend Moses Dickson were the key persons in the St. Louis relief effort. While in St. Louis, the migrants were housed mainly in three black churches: the Eighth Street Baptist Church, the Lower Baptist Church, and St. Paul's AME Church, as well as with black families across the city of St. Louis. Critical of the leadership of the Colored Relief Board, James Milton Turner tried to establish a rival organization, but it proved ineffective and ultimately turned what little money it raised over to the Colored Relief Board. It is estimated that the Colored Relief Board aided nearly six thousand exodusters.[34]

Blacks made very little progress politically in the 1870s and 1880s. Even if the Republicans had maintained control of state politics, blacks could have expected little from them. The endorsement of black political rights, which had been characteristic of the party in the late 1860s, gave way to concern about economic issues in the 1870s and 1880s. In 1878 James Milton Turner, recently returned from his seven-year stint in Liberia, sought the Republican nomination for a congressional seat from Missouri's Third District. Turner's candidacy was flatly rejected by the Republicans. So too was an effort to increase black representation on the Republican State Central Committee. Nevertheless, St. Louis was the only place in the state that blacks could expect any kind of support from the Republicans. There Chauncey I. Filley, longtime party leader and the

---

34. Nell Irvin Painter, *Exodusters: Black Migration to Kansas after Reconstruction* (New York: Knopf, 1976), 184–86, 225–30; Susanna M. Grenz, "The Exodusters of 1879: St. Louis and Kansas City Responses," *Missouri Historical Review* 73 (October 1978): 57–70. Charlton Tandy was a Civil War veteran and the captain of "Tandy's St. Louis Guard," a state militia company of black volunteers. In the 1870s other such black militia companies in St. Louis included the Attucks Guard and the Hannibal Guard (*St. Louis Missouri Republican*, October 1, 1871).

St. Louis postmaster in the 1870s, appointed a few blacks to menial patronage positions in the post office.

Black St. Louisans were able to secure more opportunity, but they failed to get an equal education. Black Missourians had been fighting segregated education for years. The 1875 Missouri constitution clearly countenanced separate educational facilities. In 1881 black leaders James Milton Turner and J. H. Murray met with Democratic Governor Thomas T. Crittenden in an effort to encourage his support of integrated schools, but Crittenden refused. In 1889 the Missouri legislature passed a law ordering separate schools to be established "for the children of African descent." Later the United States Supreme Court, in the famous *Plessy* v. *Ferguson* case, declared such separate but equal public accommodations for blacks to be constitutional. As segregation became the law of the land, facilities indeed remained separate, but never equal. Although Missouri did not pass segregation laws covering public accommodations, custom prohibited blacks from being served with whites in hotels, restaurants, theaters, and hospitals. Blacks had to enter the St. Louis Coliseum, a public building maintained by the taxes of all St. Louis citizens, by the rear door. The State of Missouri segregated a number of state facilities.

As early as 1877, St. Louis black parents had boycotted the public schools in order to force the Board of Education to hire black teachers. Black teachers were hired, and black enrollment increased in the St. Louis public schools from fifteen hundred in 1876 to thirty-six hundred in 1880. Still, in 1880 the city's eight schools for black students were in notably poorer condition than their counterparts elsewhere in the city, and five hundred black students were housed not in school structures but in rented space.[35] Black teachers were paid half as much as white teachers in 1881, and kindergartens were established in black schools only after black protests.

St. Louis reflected the urbanization of Missouri's black population during the last half of the nineteenth century. The number of blacks in the city increased sixfold between 1860 and 1880.[36] By 1890, 47 percent of the state's black population lived in cities; the figure reached a majority—55 percent—by 1900. Social, political and economic changes for blacks came in the wake of these demographic shifts. The increased concentration of blacks in the cities, and the fact that they were forced to congregate in a small area comprising a few political wards, meant that if blacks voted together, they could influence city elections. Political leaders in Kansas City and St. Louis began to take notice

---

35. David Thelen, *Paths of Resistance: Tradition and Dignity in Industrializing Missouri* (New York: Oxford University Press, 1986): 139–41; Primm, *Lion of the Valley*, 317–19.

36. Primm, *Lion of the Valley*, 314.

of this change around 1890. In St. Louis, Dr. George Bryant, a black delegate to the St. Louis Republican convention, persuaded the delegates to nominate Walter M. Farmer, a black lawyer, for the position of assistant prosecuting attorney for the Court of Criminal Correction. Another black man, W. C. Ball, was also nominated for the position of constable. Both men were defeated by Democrats. Three years later, Farmer became the first black lawyer to argue a case on behalf of a black client before the Missouri Supreme Court.

In 1892, a number of Missouri's most prominent black citizens called attention to the increased violence directed towards blacks. They hoped that educating whites to the realities of black life would make things better for them. James Milton Turner, Peter Clark, George B. Vashon, Walter Farmer, Albert Burgess, E. T. Cottman, Moses Dickson, R. H. Cole, J. H. Murray, John W. Wheeler, and a host of others circulated an address throughout the country, calling attention "to the wrongs that were being heaped upon" their fellow blacks. They called for a national day of "humiliation, fasting, and prayer" to accent their plea. The date was set for May 31, 1892, and on that day St. Louis alone saw fifteen hundred blacks solemnly gather for what they called a "lamentation day."

Despite the growing concentration of blacks in the cities, the Republican party continued to pay little attention to them. Urban Democratic leaders turned to African Americans moving into the cities. James Milton Turner had campaigned among black voters in 1888 to support the Democratic ticket. By 1890, Turner was urging black voters to leave the Republican party if it continued to ignore their interest. In 1892 former Radical William Warner ran for governor of Missouri on a "new Missouri" platform, emphasizing the state's need for industrial development and ignoring black problems. In 1896, Chauncey Filley, bastion of patronage jobs for blacks in St. Louis, fell from power, and Republican abandonment of blacks in Missouri was complete.

As the elections of 1898 rolled around, black St. Louisans contemplated a break with their old political affiliates. They expected eight hundred of the eight thousand patronage jobs available in the city. Instead, they received only seventy-six. Moreover, there were no black candidates. Internal squabbling caused perceptive Democrats to realize that the party needed new members. They quickly discovered that disaffected blacks could be wooed into the fold. Governor Lon V. Stephens wrote to several black St. Louis leaders, hoping to sway them toward the Democrats. He promised patronage jobs to blacks, said he would do what he could to have blacks appointed as policemen, and advocated a strong antilynching law.

Democrats dominated the election in black wards in Kansas City and St. Louis in 1898 and again in 1900. In St. Louis, Democratic ward boss Ed Butler and his son Jim organized black support. They were aided by a new machine organization in the city known as the Jefferson Club, which was run by Police Board Commissioner Harry B. Hawes. Through Hawes's efforts, an auxiliary Negro Jefferson Club—with C. C. Rankin, Crittenden Clark, W. H. Fields, and James Milton Turner as leaders—rallied behind Democrat Rolla Wells and helped him win the mayoral contest in St. Louis in 1901. Within a few years, however, blacks returned to the Republican party in St. Louis, because the racist elements of the Democratic party had gained control in St. Louis and statewide.

Many blacks were confined to crowded urban ghettos, where unsanitary conditions, crime, and vice prevailed. In the 1890s blacks in St. Louis lived in areas where the population density averaged eighty-two persons per acre, as opposed to the overall city average of only twelve persons per acre. Slum districts took on characters of their own and became known by such descriptive names as "Clabber Alley."[37]

Segregation caused blacks to promote the concept of self-help. A number of black businesses flourished by catering almost exclusively to a segregated clientele. There were numerous black businessmen and professionals in St. Louis, including Charles C. Clark, who ran Clark and Smith Men's Furnishing Goods Store; H. S. Ferguson, owner of the successful St. Louis Delicatessen Company; and C. K. Robinson, proprietor of the Robinson Printing Company. Yearly sales of black businesses such as these totaled more than one million dollars. Yet this impressive figure represented only about 8 percent of the estimated annual earnings of black St. Louisans. More than 90 percent of the black wage earners worked as domestics, factory workers, and common laborers.

Black newspapers of the era catered to and tried to increase black solidarity in the face of oppression. Publishers of black newspapers usually started with very limited capital. Consequently, their publications were often short-lived enterprises. One of the most successful was the *St. Louis Palladium*, founded in 1884 and edited by John W. Wheeler from 1897 to 1911. Wheeler advocated black advancement through industry and self-reliance, a philosophy eminently consistent with that of his contemporary, Booker T. Washington. Wheeler shied away from the politics of confrontation and refused to abandon the Republican party at a time when others of his race were doing so.

Perhaps one of the most positive contributions of these years of betrayal was the flowering of black music in the segregated honky-tonks and dance halls of the state. "Honest" John Turpin made St. Louis into a ragtime center. In 1880, Turpin established the Silver Dollar Saloon in the Chestnut Valley, along Chestnut and Market streets near Twentieth Street. Its best-known entertainer

---

37. Lawrence O. Christensen, "Black St. Louis: A Study in Race Relations, 1865–1916" (Ph.D. diss., University of Missouri, 1972).

"The Cascades" (sheet music cover; St. Louis: John Stark & Son, 1904). Scott Joplin was the most notable among a number of African American composers and performers who made St. Louis their home by the end of the 1800s. "The Cascades" was a musical tribute to the central feature of the 1904 Louisiana Purchase Exposition, held in St. Louis. *Missouri Historical Society*

was ragtime musician Scott Joplin, who lived for several years in Sedalia before moving to St. Louis. Joplin was born in 1868 and left home to become a pianist at age fourteen. In 1894, after composing and performing in St. Louis, he left for Chicago. When he returned to Missouri, Joplin settled in Sedalia, then a railroad center. He attended George R. Smith College, and played piano at night. In addition to his numerous celebrated rags, Joplin composed an opera, *Treemonisha,* that offered education as the key to social advancement of black people.

Other turn-of-the century black musicians in St. Louis included John Turpin's son, Tom, and W. C. Handy. It was Handy who composed the still-popular "St. Louis Blues." Tom Turpin followed in his family tradition of entertainment promotion and continued to make St. Louis a ragtime mecca. In 1897 he published his "Harlem Rag," the first published ragtime piece by a black composer. In 1899, he composed "Rag Bowery Buck," followed by "St. Louis Rag" (1903) and "Buffalo Rag" (1904). Turpin later opened the famous Rosebud Bar and the Hurrah Sporting Club. The Chestnut Valley, site of these and other clubs, inspired folktales about Stagger Lee and Frankie and Johnny. It became the base of many ragtime musicians who traveled the Mississippi Valley, seeking steady musical work in a time when white audiences preferred to hear the so-called "coon" songs from the old minstrel shows.[38]

Conclusion

Life for African American St. Louisans was in most ways better by the end of Henry Shaw's life than it had been at the time of his arrival in the city. Although turn-of-the-century St. Louis was not devoid of elements of segregation, such segregation was unequally applied. St. Louis blacks were excluded from hotels and restaurants and segregated in the city hospital, movie houses, and public amusement centers. The public library, on the other hand, was unsegregated. So too, by 1891, was the St. Louis Exposition, the city's annual industrial fair. In 1899 Walter M. Farmer, who would go on to argue before the state supreme court, became the first black man to graduate from the law department of Washington University. In 1892, two black men graduated from the university's Manual Training School. From the 1880s on, black youth attended Catholic schools. Until the second decade of the twentieth century, both black and white youths competed in track and field at the annual Public School Day.

38. George Lipsitz, *The Sidewalks of St. Louis: Places, People, and Politics in an American City* (Columbia: University of Missouri Press, 1991), 68–70; Thelen, *Paths of Resistance,* 120–29.

While St. Louis blacks started the new century enduring segregation, they were able to avoid the harsh legal measures common in southern and many other border cities.[39]

St. Louis's freedom from such laws was in large part a testament to the political power of black St. Louisans in the 1890s and early 1900s. This political power went back to the strong black community in pre–Civil War St. Louis, particularly as it coalesced in the black church and in early educational institutions. By 1891, St. Louis blacks succeeded in having their schools (once called simply "Colored School #1, #2," etc.) renamed for famous blacks. By the turn of the century, St. Louis's black and white teachers were being paid on the same salary schedule, and the expenditure per pupil was nearly equal.

At the turn of the twentieth century, most blacks in St. Louis lived in dilapidated housing in a belt running from the railroad tracks, on the south, north to Cass Avenue and west to Eighteenth Street. Some black professionals lived in better housing, particularly in the Elleardsville neighborhood (later known, as it is still, simply as "the Ville"), which lay north and west of the poorer neighborhoods. Black political leaders fought bitterly against passage of a segregated housing ordinance in 1916; the measure won approval on a citywide referendum but was ultimately rejected in the courts.

After 1900, as growing numbers of black migrants came to St. Louis from the farms and small towns of the American South, they would find an established black community with roots as deep as those of the city itself. That community supported itself with churches, schools, newspapers, health-care facilities, social recreational outlets, and employment opportunities far better than those offered in other former slave states and cities. At the dividing line of South and North, uniquely shaped by its colonial past and the simultaneous legacies of freedom and slavery, nineteenth-century St. Louis had represented an unusual, sometimes contradictory chapter in the history of urban African Americans, one that could not have taken place in any other city.

---

39. Lawrence O. Christensen, "Race Relations in St. Louis, 1865–1916," *Missouri Historical Review* 78 (January 1984): 134.

Learning from the "Majority-Minority" City
Immigration in Nineteenth-Century St. Louis

Walter D. Kamphoefner

One of the pleasures of teaching immigration history is the opportunity it affords to observe history repeating itself. The contemporary scene has been confronted with some alarmist population projections, forecasting that the middle of the new century will see a "majority-minority" society in the United States. In other words, the sum of all the ethnic and racial minorities—blacks, Asians, Hispanics, and native Americans—is predicted to constitute a majority of the population, outnumbering so-called "white Anglos."[1] There are serious grounds for skepticism about these prognostications, which are promoted mostly by two extremes of the political spectrum: on the one hand, nativists who claim that the "real America" is being undermined by waves of the wrong kind of immigration; on the other hand, self-important, largely self-appointed ethnic leaders trying to exaggerate the size and coherence of the group they supposedly represent. But even if one suspends disbelief and takes these population projections seriously, they should cause no alarm for St. Louisans. We can say, "Been there, done that." For most of the seven decades that Henry Shaw lived in the Gateway City, it was by one definition or another a majority-minority society. St. Louis's fluid social structure and heavily ethnic character facilitated the rise of newcomers such as Shaw, and such immigrants in turn played a key role in the city's rise to national prominence.

By the time Shaw first set foot there in 1819, St. Louis may well have gained a temporary Anglo-American majority, but the French still constituted an appreciable linguistic minority (even if most were by then American-born). With his school French as a foundation, Shaw had within two years of his arrival picked up what he called a "tolerable Creole French" which doubtless aided

---

1. Recent estimates have projected a "non-Latin white" proportion of 53 percent nationwide in 2050, a figure that, while still representing a majority, is significantly lower than the 69 percent recorded in the 2000 census. But "white Anglos" already constitute a minority in California. See "The New America," special issue, *Newsweek*, September 18, 2000: 40, 48.

him in business. But the French were obviously losing ground: a Federalist from Pennsylvania—William Carr Lane—defeated Auguste Chouteau in the first mayoral election of the newly incorporated city in 1823, and Chouteau's brother Pierre also ran unsuccessfully in 1826.[2] Within a couple of decades, however, native-born whites would again be reduced to a minority, and French speakers to a mere footnote, by immigration from other parts of Europe.

Shaw was certainly not St. Louis's only immigrant in 1819. The city's first newspaper had been founded by Joseph Charless, an Irish political refugee, and the Irish Catholic Mullanphy family had arrived on the scene as well.[3] There were still some natives of France among the more numerous Creoles, though most of the Francophone immigrants traced their origins to such Western Hemisphere sites as Santo Domingo and Canada. But when Julius Mallinckrodt arrived in the city in 1832, he met a French immigrant who was overjoyed to find someone, even a German, who spoke his native tongue, suggesting that the language was fading among the Creoles by then. Still, it was not until 1842 that the last French sermon was preached at the cathedral, and the last French newspaper seems to have faded from the scene in the 1850s.[4]

The Germans were one element largely missing in territorial St. Louis. One of Shaw's first business partners, the immigrant Charles Wahrendorff, is the only German mentioned in Prince Paul of Württemberg's account of his 1822 journey. While the names of two other early arrivals of prominence, Henry Geyer and Henry von Phul, sound unmistakably German, both were born in the eastern states of immigrant parentage. As in the United States generally, German immigration to the Gateway City remained insignificant before 1830, when the federal census recorded only forty-two unnaturalized foreigners in the whole

---

2. William Barnaby Faherty, *Henry Shaw: His Life and Legacies* (Columbia: University of Missouri Press, 1987), 6, 15; William E. Foley and C. David Rice, *The First Chouteaus: River Barons of Early St. Louis* (Urbana: University of Illinois Press, 1983), 196.

3. Joseph Charless, founder of the *Missouri Gazette*, was reportedly raised Methodist, but also shows up among the petitioning "members of the Roman Catholic religion" who supported the establishment of a Catholic college in 1819. Charles van Ravenswaay, *St. Louis: An Informal History of the City and its People, 1764–1865* (St. Louis: Missouri Historical Society Press, 1991), 146; Frederic L. Billon, *Annals of St. Louis in Its Territorial Days from 1804 to 1821* (1888; reprint, New York: Arno Press, 1971), 421–22; John Neal Hoover, "Joseph Charless," in *Dictionary of Missouri Biography*, ed. Lawrence O. Christensen et al. (Columbia: University of Missouri Press, 1999), 162–63.

4. Anita M. Mallinckrodt, *From Knights to Pioneers: One German Family in Westphalia and Missouri* (Carbondale: Southern Illinois University Press, 1994), 122–23; John Rothensteiner, *History of the Archdiocese of St. Louis* (St. Louis, 1928), 1:806. A French observer during the Civil War reported that his compatriots were "absolutely deprived of influence" and there was "not one single French newspaper" in St. Louis. August Laugel, *The United States during the Civil War*, ed. Allan Nevins (Bloomington: Indiana University Press, 1961), 149–50.

Sts. Peter and Paul Roman Catholic Church, Seventh and Allen Avenues, St. Louis (photograph by Emil Boehl, c. 1890). Built in the 1850s as the city's first German Catholic church, Sts. Peter and Paul provided an early sign of the growing concentration of German immigrants in South St. Louis. *Missouri Historical Society*

St. Louis population.[5] A German who arrived in 1833 wrote home that fall: "We like it here quite well . . . Sunday is kept holy here, they go to church 3 times on Sunday, and during the week besides. If there was a German church here, we would like it even better." They did not have long to wait. A couple of months later, in January 1834, the first German Evangelical congregation was founded; that same month the first German Catholic sermon was preached in the city. The first German newspaper, the *Anzeiger des Westens,* dates from 1835, suggesting that the critical mass of immigrants to sustain such institutions was just being attained.[6] But from then until the end of the century, the leading foreign element and by far the leading foreign language in St. Louis was German. How much credit is due Gottfried Duden's famous 1829 *Report on a Journey to the Western States of North America* in attracting Germans to Missouri remains open to question. The casual way Duden is mentioned in immigrant letters such as the one cited above suggests that his was a household name among the folks back home, and I know of three copies of his book that were passed down in Missouri families well into the twentieth century. However, Illinois attracted nearly 50 percent more Germans than Missouri by the Civil War without any Duden to plug for it.[7]

The stream of immigrants flowing into St. Louis after 1830 became a deluge after 1845, as the potato famine in Ireland and a milder version of the same thing in Germany—harvest failures and overall scarcity—sent hundreds of thousands of people fleeing to the New World. By 1847, as Emil Mallinckrodt noted, "one hears almost as much Low German on the streets and markets as English. . . . We live here almost as if in Germany, wholly surrounded by

5. Paul Wilhelm von Württemberg, *Reise nach dem nördlichen Amerika in den Jahren 1822 bis 1824* (1835; reprint, Munich: Borowsky, n.d.), 188; Billon, *Annals,* 265, 280, 296–97. Except for von Phul, there was no one of German descent among the thirty persons who, along with Henry Shaw, were feted in 1858 by ex-mayor John Darby for having been in business at the time of his arrival in St. Louis in 1827. John F. Darby, *Personal Recollections* (1880; reprint, New York: Arno Press, 1975), 377. St. Louis County, inclusive of the city, had a population of 14,125; the 42 unnaturalized foreigners would include, but not necessarily be restricted to, any adult males who had arrived within the previous five years. U.S. Census data compiled in *Great American History Machine* CD-ROM (University of Maryland, 1995).

6. Letter of Wilhelm and Heinrich Gerdemann, November 17, 1833, copy and transcript in Bochumer Auswandererbriefsammlung, Ruhr-University Bochum, Germany; van Ravenswaay, *St. Louis,* 303; Rothensteiner, *History,* 1:534; Karl J. R. Arndt and May E. Olson, eds., *German-American Newspapers and Periodicals, 1732–1955: History and Bibliography,* 3d ed. (Munich: Verlag Dokumentation, 1976), 250.

7. Gottfried Duden, *Report on a Journey to the Western States of North America and a Stay of Several Years along the Missouri (During the Years 1824, '25, '26, 1827),* general ed. James W. Goodrich; ed. and trans. George H. Kellner, Elsa Nagel, Adolf E. Schroeder, and W. M. Senner (Columbia: State Historical Society of Missouri and University of Missouri Press, 1980). Comparative figures on Germans in Missouri and Illinois can be found in *Eighth Census of the United States, 1860. Population* (Washington, D.C., 1864), 620–21.

Germans."[8] The next year, a failed revolution in Germany would add a leavening of intellectuals and political radicals to the mixture, along with more ordinary immigrants than ever before.

The 1840s saw the most dramatic decade of growth ever in St. Louis: the city nearly quadrupled. The association with the immigrant influx was more than coincidental. The 1850 census, the first to give a tally of birthplaces, revealed that the majority of St. Louisans had been born abroad. Natives of Germany, comprising nearly one-third of the population, outnumbered natives of Missouri. Holding second place among immigrants were the Irish, whom Germans outnumbered by more than two to one. Henry Shaw and his British countrymen were just a small minority: less than 5 percent of St. Louisans in 1850 and going downhill from there. The French were the next largest group, and the largest foreign-language group after Germans, but they never claimed more three thousand persons, or 2 percent of the city's inhabitants.

Readers who have studied the history of St. Louis may have read that it led the nation's cities in 1860 in the proportion of foreign-born, 60 percent of the population. In fact, this figure, drawn from a census introduction, is wrong—even if many good historians (including myself in my first publication) have trusted it. It was arrived at by dividing the foreign-born population of the city and county by the total population of just the city; a more detailed breakdown, appearing on another page of the same publication, makes the source of the error obvious. The correct figure of just over 50 percent is still quite impressive, tying the Gateway City with Chicago, San Francisco, and Milwaukee for the national lead.[9]

---

8. Mallinckrodt, *Knights to Pioneers*, 245.

9. Erroneous figures are derived from *Eighth Census*, xxxi–xxxii, with correct figures for the entire county on 297, 300, 614. The 1860 countywide figures may somewhat underestimate the foreign-born proportion in the city, which according to the 1858 city census surpassed 56 percent, with an Irish share of 16 percent and a German share of 32 percent or almost one-third (*Anzeiger des Westens*, October 24, 1858). Readers interested in exploring nineteenth-century St. Louis population statistics will find additional information in *Statistical View of the United States, . . . a Compendium of the Seventh Census* (Washington, D.C., 1854), 399; *Ninth Census, Volume 1. The Statistics of Population* (Washington, D.C., 1872), 110, 194, 380–89; *Statistics of the United States at the Tenth Census, 1880. Volume 1, Population* (Washington, D.C., 1883), 518, 538–39; *Compendium of the Eleventh Census, 1890. Part 2* (Washington, D.C., 1892), 600–11. The mistaken 1860 figures can be found in Selwyn K. Troen and Glenn E. Holt, eds., *St. Louis* (New York: New Viewpoints, 1977), xxi; Selwyn K. Troen, *The Public and the Schools: Shaping the St. Louis System, 1838–1920* (Columbia: University of Missouri Press, 1975), 58; Martin G. Towey, "Kerry Patch Revisited: Irish Americans in St. Louis in the Turn of the Century Era," in *From Paddy to Studs: Irish-American Communities in the Turn of the Century Era, 1880 to 1920*, ed. Timothy J. Meagher (New York: Greenwood Press, 1986), 142; Phillip T. Tucker, *The Confederacy's Fighting Chaplain: Father John B. Bannon* (Tuscaloosa: University of Alabama Press, 1992), 12; Walter D. Kamphoefner, "St. Louis Germans and the Republican Party, 1848–1860," *Mid-America* 57 (1975): 70.

The Civil War slowed immigration nationwide, but St. Louis was hit especially hard. Its Mississippi River connection to New Orleans, the city that one historian has dubbed the "back door to the land of plenty," was blocked until summer 1863, whereas rival Chicago kept open its East Coast connections and received further stimulus as the jumping-off point for the transcontinental railroad. St. Louis maintained its population lead in 1870 by a hair (and a bit of census fraud—about which, see James Neal Primm's essay in this book). But not even the great fire of 1871 could stop Chicago from vaulting into the lead by 1880, while a disbelieving St. Louis demanded a census recount that did nothing but confirm its demotion.[10] Much of the difference in growth rates depended on the continued ability to attract immigrants. Chicago, too, lost its immigrant majority in 1870, but just barely. Though still more than ten thousand behind St. Louis in total population at that time, Chicago's immigrants—48 percent of the city's total population—already outnumbered St. Louis's by thirty-two thousand.

Immigrants made up only 36 percent of the St. Louis population in 1870, though its black percentage had jumped from 2 to 7 percent in the decade of emancipation. While the two add up to only 43 percent, in another sense the Gateway City was still by far a "majority-minority" society, and would continue to be for the rest of the century. Many whom the census designated as "Americans" were in fact the children of German and Irish immigrants. More than half of the city's native-born had immigrant parents, while old-stock Americans accounted for less than one-third of the city's population. In fact, taking the second generation into account, St. Louis came close to having a German majority. This is quite apparent from the 1880 school population. While only 6 percent of all white schoolchildren in St. Louis were foreign-born, barely one-quarter of their parents had been born in the United States; nearly half of their parents (46 percent, to be exact) were natives of Germany.[11] As late as the turn of the century, native whites of native parentage still constituted less then one-third of the city's population.

In terms of urban geography, there was not one single German neighborhood, as in many other cities, but instead heavy concentrations on both the north and south sides, with few in the middle wards. At least one-third of the city's Germans lived in relative isolation on the South Side, where English speakers comprised only a small minority and where ready access to the rest of the city was impeded by the belt of railroad tracks extending through the Mill

---

10. Frederick M. Spletstoser, "Back Door to the Land of Plenty: New Orleans as an Immigrant Port, 1820–1860" (Ph.D. diss., Louisiana State University, 1978); National Archives and Records Service, *Federal Population Censuses, 1798–1880* (Washington, D.C., 1978), 66.

11. Troen, *Public and the Schools,* 58.

Creek Valley. Some of my earlier research has analyzed the 1860 census data on the Second Ward, a long, narrow strip running from east to west and spanning the social scale from the rough levee district and the literally floating population living on steamboats to the fashionable neighborhoods around Lafayette Park housing the likes of James B. Eads, whose fortune was listed at four hundred thousand dollars. Eads was among the small minority of Anglo-Americans in his ward, which formed part of the heavily German South Side of St. Louis. About two-thirds of the ward's inhabitants over age twenty-one had been born in the German Empire; to their number was added a substantial population of German speakers from Switzerland, Alsace, and the Austro-Hungarian Empire. Irish, making up 7 percent, were the largest immigrant group besides Germans, and together with a scattering from Britain and Canada they brought the proportion of Anglophone immigrants up to 10 percent. Out of an adult population of nearly seven thousand, there were barely six hundred born in the United States, so native speakers of English were outnumbered almost four to one among adults in this ward. The adjacent First Ward, to the south, was almost identical in composition.[12]

One index of the social distance between Germans and Irish was their low intermarriage rate. Married couples of the Second Ward in 1860 included over fifteen hundred German men and nearly fourteen hundred German women who had no children born outside the United States and presumably married here. But despite a sizeable group from Catholic source areas, not a single German woman was married to an Irishman, and only eight German men—just half of 1 percent of their total number—had taken Irish brides. This pattern extended to later decades and across the city and state. The 1880 census shows that of all the children born in Missouri to at least one German parent, less than 1 percent also had an Irish parent. Since the Irish are a smaller group, the same number of German-Irish intermarriages produces a higher percentage among their population. Even so, less than 1 percent of Missouri children with Irish fathers had German mothers, and children with German fathers made up only 2 percent of all children with Irish mothers.[13] Literally and figuratively, North St. Louis's Kerry Patch was a long way from the South Side.

Immigrant letters give further evidence of the self-contained world in which St. Louis Germans could live if they chose. One immigrant wrote home to the Fatherland with respect to his south-side wife who was born in 1870: "Why do

---

12. Walter D. Kamphoefner, "Paths of Urbanization: St. Louis in 1860," in *Emigration and Settlement Patterns of German Communities in North America*, ed. E. Reichmann, L. Rippley, and J. Nagler (Indianapolis: Max Kade German-American Center, 1995), 259.

13. Calculated from ibid., 264, and *Statistics of the United States at the Tenth Census, 1880*, 686. In 1880, over half of all Missouri Germans and 58 percent of the state's Irish lived in St. Louis.

you keep asking if my wife is German? Of course she was born here, but she speaks German just as well as any of you and is proud of her German heritage." However, he did indicate that she couldn't write the language, claiming erroneously that the public schools did not teach it. As a matter of fact, in the 1870s and 1880s, the American-born daughter of Jetta Bruns supported herself as a young widow doing just that, teaching German in the public schools. As her mother related, when German instruction was abolished, teachers were given a chance to qualify for other school jobs by taking an English exam. "However, Effie has become so timid, and although she had spent a great deal of effort and expense earlier in preparing herself, she now explains that she has become too old, etc., etc. For the time being she is giving private lessons to German children in school." As far as can be determined from her mother's letters, she never did qualify herself to teach in English, despite the fact that she was a doctor's daughter and had been born and raised in Missouri. In fact, only twenty-eight of ninety-eight former German teachers were able to qualify for reassignment.[14]

One of the assumptions behind the "majority-minority" thesis, at least among the fearmongers in the "English-only" crowd, is that minorities will unite in common cause against the "responsible" elements of the population—specifically, the outnumbered Anglos. Expressed conversely, in the imagery of self-appointed ethnic leaders, this assumption holds that various "oppressed" elements will rise up in solidarity with one another. Here too, the operation of the "majority-minority" society in nineteenth-century St. Louis offers some insights. One thing quickly becomes apparent: as intermarriage rates have already suggested, various ethnic groups were as much at odds with one another as with Anglo-Americans, who were themselves far from being unified.

The reigning interpretation of nineteenth-century American politics depends on an assumption of ethnocultural polarization. According to this view, the Whig party and its Republican successor were dominated by crusading Protestant do-gooders bent on improving society and individuals, whether they realized they needed improving or not. Arrayed against them in the Democratic party were various groups who wanted to be left alone in their own subcultures: not only southern slaveholders and white supremacists, but most immigrants and Catholics as well. But given the city's French-Catholic roots, the political landscape of St. Louis contrasted rather sharply with this model. One of the

---

14. Walter D. Kamphoefner, Wolfgang Helbich, and Ulrike Sommer, eds., *News from the Land of Freedom: German Immigrants Write Home* (Ithaca: Cornell University Press, 1991), 495; Adolf E. Schroeder, ed. and trans., *Hold Dear, As Always: Jetta, A German Immigrant Life in Letters* (Columbia: University of Missouri Press, 1988), 230, 264 ff.; Audrey Olson, "St. Louis Germans, 1850–1920: The Nature of an Immigrant Community and its Relation to the Assimilation Process" (Ph.D. diss., University of Kansas, 1970), 106.

leading Whig elements was the old French Creole elite, which restricted the anti-Catholic and antiforeign forces to an undercurrent that only occasionally boiled up to the surface. From the 1830s on, the Democratic party was supported by an immigrant coalition of Irish and Germans, a cooperation that grew increasingly shaky in the 1850s and collapsed entirely by 1860. But a successor to the Whig party persisted until the Civil War; the Republicans, rather than springing up from Whig ashes as they did elsewhere, arose in St. Louis out of a Democratic split between that party's proslavery and Free-Soil wings, whereas many conservative former Whigs became Democrats. This odd party evolution was reflected in the St. Louis press: during and after the Civil War, a leading Democratic paper was called the *Republican,* harking back to its prewar "National Republican" origins—an old name for the Whigs. Manifesting its Free-Soil Democratic roots, the leading Republican paper was called the *Democrat*—as it was still when I was growing up in Missouri.[15]

Political, ethnic, and cultural realms are, as the selection of essays in this book suggests, so closely intertwined as to make the separation of one from the other impractical, if not impossible. Conflict at the polls and conflict in the streets were often closely related to one another. The closest the Gateway City came to an immigrant-native polarization was in the two decades before the Civil War. The tremendous immigrant influx into St. Louis did not go unnoticed, or unchallenged. The first recorded election riot was in 1844, by which time an active "Native American" movement aimed to keep immigrants and Catholics out of public office, and to enact a twenty-one-year waiting period before immigrants could become naturalized. Nativists managed to elect Peter Camden as mayor in 1846, but were largely discredited when he imposed his notions of Anglo-Protestant Christianity to the extent of stopping omnibus service on Sunday afternoons.[16]

Nationally and locally, the nativist movement peaked in the early 1850s when the breakdown of the existing party system coincided with the heaviest influx of immigrants the nation has ever experienced. There were only about twenty years in all U.S. history when immigration exceeded 1 percent of the current population, but eight of them came in succession, starting in 1847 and peaking with an influx of nearly 1.8 percent in 1854. (Perhaps not coincidentally, 1854 was the year the value of St. Louis beer production surpassed one million dollars.) With the number of inhabitants and foreign-born doubling during the

---

15. Walter D. Kamphoefner, "German-Americans and Civil War Politics: A Reconsideration of the Ethnocultural Thesis," *Civil War History* 37 (1991): 226–46; Kamphoefner, "St. Louis Germans," 69–88; Galusha Anderson, *The Story of a Border City during the Civil War* (Boston: Little, Brown, 1908), 143–44.
16. Towey, "Kerry Patch Revisited," 143.

1850s, the impact in St. Louis was all the more dramatic—as were the reactions. An 1849 confrontation between firemen and heckling Irish bystanders, for example, escalated into a riot that wrecked several Irish establishments, but, although a cannon was brought into position, no lives were lost.[17]

The next years would not be so lucky. In the April 1852 city election, rumors (probably true) that Germans were hindering Whig voters on the South Side sent a nativist mob allegedly numbering several thousand heading south bent on revenge, led by professional rabble-rouser Ned Buntline. After storming the polling place and destroying the ballot box, the mob turned on a nearby German saloon. One of the leaders tried to break down the door and was shot and killed by the saloon keeper, whereupon the mob set the saloon afire and burned it to the ground, with the saloon keeper, his wife, and their newborn infant barely escaping with their lives. After dark, the rabble descended on the offices of the leading German newspaper, the *Anzeiger*, but were prevented by a militia company from doing any damage.[18] Two years later, in August 1854, a polling-place fight escalated into a major riot in the Irish Fifth Ward, requiring two days to suppress and leaving ten persons dead, thirty-three wounded, and nearly one hundred buildings damaged.[19]

Against this background, one might think that Germans and Irish would have united against the common enemy of nativism, but as Primm puts it, there was "little love lost" between the two. A St. Charles German paper commented after the 1854 violence: "We are very happy . . . that the Germans took no part in the riot itself. They acted more circumspectly than the Irish did two years ago, when they made common cause with the Americans against the Germans. Truly, the immigrants here have no reason to feud with each other, but rather to unite for the protection of their lives and property."[20]

As it turned out, the institution of a professional, uniformed police force helped prevent any additional riots during the decade. But the Irish and Germans, already squabbling within the Democratic party over the ethnic balance on the ticket, from the mid-1850s on increasingly went their separate ways. In the three-way political races leading up to the Civil War, the Free-Soil

---

17. Brinley Thomas, *Migration and Economic Growth: A Study of Great Britain and the Atlantic Economy*, 2d ed. (Cambridge: Cambridge University Press, 1973), 443; Charles van Ravenswaay, "Years of Turmoil, Years of Growth: St. Louis in the 1850's," *Bulletin of the Missouri Historical Society* 23 (1967): 309; Primm, *Lion of the Valley*, 166–67.

18. Primm, *Lion of the Valley*, 168–69; Heinrich Börnstein, *Fünfundsiebzig Jahre in der Alten und Neuen Welt* (1881; reprint, New York: Peter Lang, 1986), 2:152–57; Jay Monaghan, *Great Rascal: The Life and Adventures of Ned Buntline* (Boston: Little, Brown, 1951), 197–203.

19. John G. Schneider, "Riot and Reaction in St. Louis, 1854–1856," *Missouri Historical Review* 68 (1974): 171–75; Primm, *Lion of the Valley*, 170–72; van Ravenswaay, "Years of Turmoil," 311–12.

20. Primm, *Lion of the Valley*, 169; *St. Charles Demokrat*, August 26, 1854.

Benton Democrats and their Republican heirs drew their prime support from Germans, the "National" (but increasingly pro-Southern) Democrats from the Irish, and the Whig party's nativist successors from Americans. As late as 1880, a newspaper editorial characterized St. Louis as "still divided, like all Gaul into three parts, disputed by Ireland, Germany, and aboriginal America."[21]

Martin G. Towey argues convincingly that the Irish experienced less prejudice and discrimination in St. Louis than in the Northeast. After all, what other city could claim as its first millionaire an Irish Catholic like John Mullanphy? Among the contributing factors, Towey points first of all to the Catholic character of the city from its founding. Secondly, he argues, the adamantly Unionist Germans served as a lightning rod for prejudice, especially on the part of native Southern sympathizers.[22]

Although Confederate sympathizers comprised a significant element of St. Louis's population, they did not set the dominant tone, even among the city's native-born. In terms of its population, St. Louis remained a Northern city in a Southern state. While the slave states of Kentucky, Tennessee, and Virginia were the leading sources of migration to Missouri overall, the states that had contributed the most migrants to St. Louis by 1860 were New York, Pennsylvania, Ohio, and Illinois, in that order. The city looked even less Southern with respect to its black population, which hit a low point of 2 percent on the eve of the Civil War—more like New York than New Orleans.

Even if many of the city's Anglos were cautious Unionists, St. Louis Germans were the most enthusiastic supporters of the Union and emancipation. The first shots of the Civil War in Missouri were fired on the streets of St. Louis on May 10, 1861, in a riot with strong ethnic overtones. "Wide-Awake" German militia companies had been drilling since the election and, in cooperation with several Anglo Unionist politicians, played a decisive role in preventing the federal arsenal from falling into Confederate hands. So far, so good, but when they captured the secessionist state militia at Camp Jackson and marched them through the streets, it triggered a riot that left twenty-eight dead. Although this no doubt pushed some fence-sitters over to the Confederate side, Germans provided much of the manpower (nine-tenths of the five three-month regiments raised at the war's outbreak) that quickly put the Union in control of this crucial border state.[23]

---

21. On voting patterns in the decade leading up to the war, see Kamphoefner, "St. Louis Germans." The newspaper editorialist's "aboriginal America"—from Towey, "Kerry Patch Revisited," 139—of course does not refer to Indians.

22. Towey, "Kerry Patch Revisited," 141–42.

23. For a good overview of the coming of the Civil War in the city, see *Germans for a Free Missouri: Translations from the St. Louis Radical Press, 1857–1862*, selected and translated by Steven

Missouri Germans demonstrated a high level of Union army participation compared not only to other Missourians, but also to other Germans nationwide. Though Missouri had only the sixth largest German population in 1860, it was second only to New York State in the number of Germans who rallied to the blue. While there were obviously St. Louis Irish who served the Union and probably some Germans who fought for the Confederacy, Captain Joseph Kelly's Irish company had no German equivalent among Missouri's rebel troops. And there was certainly no German counterpart to Father John B. Bannon, a Dublin-born St. Louis priest who became the "Fighting Chaplain" of a Missouri rebel brigade and then went on to plead the case for Confederate diplomatic recognition before the pope.[24]

Witnesses of various ethnic backgrounds attested to the German devotion to the Union cause. Abolitionist Baptist preacher Galusha Anderson may have had reservations about their beer drinking, but he stated that "the Germans of the city . . . with hardly an exception were openly and stoutly opposed to secession." Indeed, beer drinking and Unionism seem to have gone together, as an Anglo soldier reported after returning to St. Louis from the 1861 southwest Missouri campaign:

> I never put my head out of the hotel but that—having on my First Iowa uniform—the first German who saw me took me by the arm to the nearest beer saloon, and . . . said, 'You fights mit Sigel—you drinks mit me.' . . . . The Germans were fervently and joyfully patriotic; they could not do enough for anyone who had 'fought mit Sigel.' The intense rebel element in St. Louis was still alive and active, but it was driven entirely out of sight by the intensity and vigor of the enthusiastic patriotism of the Germans.

Despite German impatience with the Lincoln administration, their fervor continued throughout the war, as a French traveler observed in 1864: "It is perhaps amongst the German population of St. Louis that we must look for the most ardent defenders of the Union, and the most resolute enemies of slavery."[25]

Rowan, with introduction and commentary by James Neal Primm (Columbia: University of Missouri Press, 1983), and Louis Gerteis, *Civil War St. Louis* (Lawrence: University of Kansas Press, 2001). Greater detail is provided by van Ravensway, *St. Louis*, 458–551. Anderson, *Story of a Border City*, is an excellent eyewitness account.

24. Kamphoefner, "German-Americans and Civil War Politics," 244–46; Phillip T. Tucker, "John B. Bannon," in Christensen et al., eds., *Dictionary of Missouri Biography*, 29–30. Tucker, *Fighting Chaplain*, 13–21, indicates that a majority of St. Louis Irish served the Union, if with mixed feelings. Sentiment at the top was more pro-Confederate; Archbishop Peter Kenrick refused to fly Union flags over the city's Catholic churches. See Rowan and Primm, eds., *Germans for a Free Missouri*, 224–29.

25. Anderson, *Story of a Border City*, 160, 311–12; E. F. Ware, *The Lyon Campaign in Missouri, Being a History of the First Iowa Infantry* (1907; reprint, Iowa City: Camp Pope Bookshop, 1991), 347; Laugel, *United States*, 150.

The "extreme bitterness" against St. Louis Germans held by many southern sympathizers is beyond dispute; as the Reverend Anderson delicately observed, "[M]ore than once I heard them hotly denounce the Germans as Amsterdam Dutch without the Amster." Nevertheless, it was during and immediately after the war that the Germans reached the height of their influence in city and state politics, even electing one of their own, Forty-eighter Carl Schurz, as senator from Missouri in 1869. And despite the detour of the Liberal Republican interlude, Germans continued to be the bulwark of Republicanism for the rest of the century. Needless to say, they were at odds with the Irish from then on.[26]

Comparing the relative political success and influence of the two groups proves difficult. Already during the 1840s, St. Louis had elected three Irish mayors, two of them immigrants and at least two Catholics.[27] German immigrant Henry Kayser had served as appointive city engineer under ten different mayors between 1839 and 1857, interrupted only when nativists held city hall.[28] But it was not until 1853, when Henry Overstolz became controller, that a German captured a citywide elective office. Overstolz went on to become the city's first German mayor in 1876, and in 1877 was elected to a four-year term under the new city charter. So the city actually spent more time under this one German mayor than under the three Irish mayors who served during Henry Shaw's lifetime (though a second-generation Irish Catholic, Edward Noonan, was elected in 1889). Beyond formal elective office, one should mention the informal power exercised by Irish immigrant boss Ed Butler, even if it was not as unlimited as some observers have imagined.[29] Though outnumbered more than two to one, the Irish held nearly as many lower-level offices as Germans in the post–Civil War era, a consequence of the language factor as well as greater ethnic unity. This also made itself felt in their heavy representation on the police force.[30]

---

26. Anderson, *Story of a Border City*, 160. Hans L. Trefousse, *Carl Schurz: A Biography* (Knoxville: University of Tennessee Press, 1982), 170–81.

27. Mayor James Barry was a Catholic immigrant and Bryan Mullanphy the son of one. Immigrant George Maguire's religion is unknown, but he may well have been Catholic given that his wife was of the French Provenchere family. Melvin Holli and Peter d'A. Jones, eds., *Biographical Dictionary of American Mayors, 1820 to 1980* (Westport, Conn.: Greenwood Press, 1981), 18, 240, 266, 418–31.

28. Gustav Körner, *Das Deutsche Element in den Vereinigten Staaten von Nordamerika, 1818–1848* (1880; reprint, New York: Peter Lang, 1986), 308–10; Eric Sandweiss, "Henry Kayser," in Christensen et al., eds., *Dictionary of Missouri Biography*, 446.

29. "St. Louis's supposed boss Ed Butler was more of a freelance political speculator who bargained for favors than a boss who dictated policy," writes Jon C. Teaford in *The Unheralded Triumph: City Government in America, 1870–1900* (Baltimore: Johns Hopkins University Press, 1984), 52; see also 179–80. Cf. Edward C. Rafferty, "Edward Butler," in Christensen et al., eds., *Dictionary of Missouri Biography*, 140–42.

30. Towey, "Kerry Patch Revisited," 149–53.

Various historians have emphasized the disunity of St. Louis Germans, perhaps more so than is warranted.[31] Certainly Catholics were less inclined to vote Republican than other Germans; one of them, immigrant Henry J. Spaunhorst, served three terms as a Democratic state senator and was the only Catholic at the State Constitutional Convention of 1875. But in 1892, three years after Henry Shaw's death, the city would send a German-born Republican, Richard Bartholdt, "from steerage to Congress" (as he put it in his autobiography) and keep him there for twenty years.[32] With respect to the mayoral races in the nineteenth and twentieth centuries, St. Louis Germans did quite well compared to the Irish or to their countrymen in other cities. During the time span from 1823 through 1980, German immigrants or their children presided over city hall for twenty-five years, four times as long as Irish Catholics and longer than in such heavily German cities as Milwaukee and Cincinnati. Most of these mayors served after Henry Shaw had passed from the scene, but their election was still a legacy of the immigration of his era. Including those of the third generation like Henry Kiel, six German mayors through 1948 were Republican; there were just two German Democrats, one during the New Deal and another in 1848 before there were was a Republican party.[33] Even allowing for multiethnic coalitions, this suggests that German disunity has often been exaggerated.

African Americans and Germans were political allies in St. Louis for decades after the Civil War, but the exact nature of their relationship was complicated and sometimes difficult to discern.[34] Before emancipation, Germans had little to do with slavery—or with blacks in any capacity, slave or free. The heavily German First Ward in 1850 had the most white inhabitants, 19 percent of the citywide total, but only 3 percent of all slaves and just 2 percent of all free blacks in the city. Throughout the Civil War, St. Louis Germans and their rural countrymen adamantly lobbied Lincoln for emancipation, and many were ready to desert him for the more radical John Fremont in the 1864 presidential election. A St. Charles German presided over the convention when Missouri abolished slavery in 1865, with strong support from his St. Louis compatriots. At the beginning of Reconstruction, south St. Louis legislator Gustav Finkelburg was the leading force in support of black voting rights before that

31. In particular, Olson, "St. Louis Germans," 19–36 passim. A less detailed summary of her dissertation was published as "The Nature of an Immigrant Community: St. Louis Germans, 1850–1920," *Missouri Historical Review* 66 (1972): 342–59.

32. Richard Bartholdt, *From Steerage to Congress: Reminiscences and Reflections* (Philadelphia: Dorrance, 1930). For Spaunhorst, see Rothensteiner, *History*, 2:414.

33. Tallied from information in Holli and Jones, eds., *Biographical Dictionary of American Mayors*.

34. Towey, "Kerry Patch Revisited," 151.

issue was placed before the people as a proposed constitutional amendment in the 1868 election. Carl Schurz and other leaders in the St. Louis German press strongly supported the amendment, but the rank and file apparently had other opinions.[35] In the presidential race Grant carried the city easily, with better than two-to-one margins in the three south-side German wards. But only 45 percent of those who supported Grant also voted for black suffrage, and the cause fared no better on the South Side than elsewhere. In fact, the "no" vote was greater there than in the rest of the city compared to the number of Democratic votes for president. "No rose without thorns" was the reaction of the *Westliche Post,* which went on to comment that "right remains right, whether or not it achieves victory on the first try."[36] Still, after the detour into the Liberal Republican movement in 1872, St. Louis Germans largely returned to the Republican fold, although their prewar native standard-bearers—including Frank Blair and B. Gratz Brown—ended their careers as Democrats. Even if German racial idealism faded somewhat in later decades, it remained strong enough in 1916 that Republican Mayor Henry Kiel opposed a referendum to pass a racial segregation ordinance, and German socialists supplied most of what few white votes were cast against it.[37]

It would be mistaken to attribute contrasts between Germans and Irish in political affiliation and racial attitudes simply to German virtue or the humanitarian ethos preached in their liberal press. In their neighborhoods and on the job, Germans were much less likely than Irish to encounter African Americans. Coming from a more developed economy, they were often able to fill niches as skilled artisans, or to work for their better-off countrymen in industries such as brewing, which they dominated. Similarly, the bulk of German women who worked as domestics served their own compatriots rather than Anglo-Americans, and thus encountered little competition from blacks. The Irish, by contrast, were overwhelmingly rural in background, arrived with few marketable skills, and often started out at the bottom, competing directly with blacks. As an observer noted in 1874, "The Hibernian is also prominent in

---

35. Donald Scott Barton, "Divided Houses: The Civil War Party System in the Border States" (Ph.D. diss., Texas A&M University, 1991), 185–98. Finkelburg's residence in the south-side Third Ward is documented by the *U.S. Census Manuscripts, 1870,* Roll 811, p. 419.

36. *Westliche Post,* November 5, 1868; election returns were published on November 6. Throughout its existence, the *Post* continued to be an "unequivocal supporter of civil rights, especially . . . of German-Americans and Negroes." Harvey Saalberg, "The *Westliche Post* of St Louis: A Daily Newspaper for German-Americans, 1857–1938" (Ph.D. diss., University of Missouri, 1967), 415.

37. Primm, *Lion of the Valley,* 411–13; *St. Louis Labor,* March 4, 1916, quoted in Troen and Holt, eds., *St. Louis,* 84. While the referendum passed, the segregation ordinance was later struck down in the courts.

St. Louis; he has crept into the hotel service, and the [N]egro has sought another field of occupation."[38]

Although the Germans and Irish far outnumbered any other groups in St. Louis, they were not the only groups large enough to support institutions of their own. The first Jewish congregations (themselves primarily consisting of Germans) were established in the 1840s, with ground being broken for the first synagogue in 1855. *Anzeiger* editor Henry Boernstein's 1851 anti-Jesuit potboiler was translated not only into English (as *The Mysteries of St. Louis*) but also into French and Czech for local newspaper serialization.[39] In 1854 St. John Nepomuk was established as the nation's first Bohemian Catholic parish. St. Louis also had the first Bohemian benefit society nationwide, and just missed having the first Czech-language newspaper. Besides the German refugees of the 1848 revolution, there were others from the Austro-Hungarian Empire (primarily Hungarians and Jews), though many came speaking German as a second if not a first language, and often integrated more quickly into that community than into Anglo society. Most, for example, served in German units during the Civil War—Joseph Pulitzer being a good example. Occasionally, someone such as Isidor Bush was elected to political office in a German district. By the Civil War, ten different foreign languages were spoken in the city according to the school superintendent, but German was by far the most widespread.

As of 1874, St. Louis supported four daily newspapers in the German language alone, alongside five English papers. Three of the German dailies outlived Henry Shaw, and on a weekly or monthly basis there were numerous other publications ranging across the political spectrum, including several anarchist and socialist sheets. One of the latter, the *Arbeiter Zeitung*, gave rise to an English counterpart, *St. Louis Labor*, but as late as 1922 over 40 percent of their combined circulation was in German. Both the Lutherans and the Evangelicals published German religious monthlies that lasted for more than a century, spanning Shaw's lifetime and my own (though with parallel English publications in their latter decades). St. Louis Catholics were able to support a German-language daily down through World War I.[40]

---

38. Kamphoefner, "Paths of Urbanization," 263; in the Second Ward in 1860, only one-eighth of all German domestics worked for Americans, although the proportion was probably higher in more fashionable Anglo wards. Towey, "Kerry Patch Revisited," 145–49; Edward King, *The Great South* (1875; reprint, New York: Arno Press, 1969), 225; also excerpted by Troen and Holt, *St. Louis*, 73.

39. Walter Ehrlich, *Zion in the Valley: The Jewish Community of St. Louis* (Columbia: University of Missouri Press, 1997), 1:88–103; Steven Rowan, "Henry Boernstein," in Christensen et al., eds., *Dictionary of Missouri Biography*, 89.

40. Arndt and Olson, *German-American Newspapers*, 258–59, 262–63. The century-spanning monthlies were *Der Lutheraner*, 1844–1954, and *Der Friedensbote*, 1850–1955.

Largely forgotten in today's debates over the public use of languages other than English (particularly in areas with large Hispanic populations) is the extent to which nineteenth-century Americans met immigrants halfway. The case of St. Louis illustrates this fact well. Beginning in 1865, when the influence of St. Louis Germans in state affairs was at its height, Missouri started allowing immigrants to vote after only one year of residence, provided they had taken out "first papers" declaring their intent to become citizens.[41] Ironically, it was under Irish-born mayor James Barry that St. Louis in 1849 arranged to have all city ordinances translated into German, but this is perhaps less surprising given that the two groups were political allies at the time.[42]

The most outstanding example of accommodating immigrants was the program offering German instruction in St. Louis public elementary schools, but here the Irish were among the Germans' most bitter opponents. The program was instituted in 1864 after years of agitation, and at a time when German Republicans were at the height of their influence and some Confederate sympathizers were excluded from voting. Begun as an experiment in five schools the first year, German instruction met with heavy demand and quickly spread throughout the city's system. Within ten years, nearly half of all pupils were taking part; in 1878, the *Westliche Post* proudly reported a figure of 53 percent. By 1880, fifty-two of the city's fifty-seven public schools offered the program. In another respect as well, Anglo-Americans exhibited more open-mindedness than historians have given them credit for: nearly one-quarter of the pupils taking language instruction, some five to six thousand in a typical year, were not of German origin. German instruction was offered as just one of the courses during a normal day of instruction, so it did not have the effect of isolating German students from others as was the case with the half-German, half-English programs in other cities such as Cincinnati.[43]

Although such programs in some other cities lasted until World War I, St. Louis's was abolished in 1887, in some respects a victim of its own success. German-language instruction added an estimated one hundred thousand dollars annually to the school budget. In 1878 an Irish board member introduced a petition to teach Gaelic, expecting that it would be denied and thus present a justification for dropping German on the grounds of equal treatment. A group of Germans organized to fight the proposal, appointing Reform Rabbi Solomon Sonnenschein to present their case before the school board. (One wonders

---

41. Walter B. Stevens, *Centennial History of Missouri* (St. Louis and Chicago: S. J. Clarke Publishing, 1921), 1:877.
42. Holli and Jones, eds., *Biographical Dictionary of American Mayors*, 18.
43. Troen, *Public and the Schools*, 55–65; *Westliche Post*, January 24, 1878; Kamphoefner, Helbich, and Sommer, *News from the Land of Freedom*, 22.

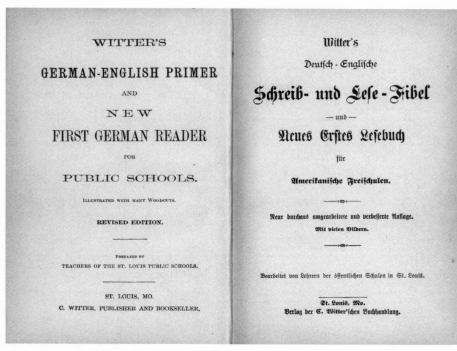

Title pages in English and German for *Witter's German-English Primer and New First German Reader for Public Schools* (St. Louis: C. Witter, 1881). German language instruction was a regular feature of St. Louis public education from its introduction in 1864 until 1887. *Missouri Historical Society*

whether this appointment reflects the German community's openness toward Jews, the rabbi's personal reputation, or perhaps an effort to avoid making either Catholics or Lutherans feel that the other group was benefiting from favoritism). Within a month, presented with forty thousand signatures favoring retention, the school board relented. It is of course impossible to determine what proportion of the petition's signatories were German, but we can safely presume that it was significant. The overall number of signatures—equal to 73 percent of all German-born men, women, and children in the 1880 census—suggests that one can hardly claim, as one scholar did, that Germans were indifferent to language instruction. In the end, it took the state legislature to eliminate it from the public schools. In 1887, in what historian Selwyn Troen has termed "a clear case of gerrymandering," legislators redrew school board districts in such a way as to lump as many Germans as possible together in one district and distribute the rest so that they would have no chance of forming majorities. This, combined with Republican division and disorganization, left the

German "Turnverein" ticket with only three of twenty-one school board seats. The fate of German instruction was sealed. The Germans' reaction to cancellation can best be described as resignation. Few forsook the public schools on that account, though down to World War I, Saturday language instruction supported by voluntary contributions drew enrollments of over one thousand.[44]

What lessons can one draw from this chapter of St. Louis history? An Anglo observer remarked in 1874: "The city acted wisely in introducing the study of German, as otherwise the Teutonic citizen would doubtless have been tempted to send his child to a private school. . . . Now native American children take up German reading and oral lessons at the same time as their little German fellow-scholars." Troen confirms the integrative effects of this program. Whereas four-fifths of German children attended parochial or private schools before 1864, by 1887 probably 80 percent were in the public system.[45] At least in St. Louis, Yankee Protestant Republicans proved to be more pluralistic than their reputation. Connecticut-born William Torrey Harris (about whom see William J. Reese's essay in this volume), with his Puritan roots and Yale education, would hardly have seemed predisposed to show sympathy for German immigrant culture. Yet it was during his tenure as superintendent from 1868 to 1880, and with his support, that German language instruction in St. Louis schools reached its apex.

There were other manifestations of Anglo-German cross-fertilization in the intellectual circle around Superintendent Harris. Susan Blow's advocacy of the kindergarten movement, in the St. Louis school system and the nation generally (also discussed in Reese's essay), derived from ideas she had been exposed to while touring Germany and from her experience with immigrants in the United States.[46] Harris, Blow, and several young Germans formed the core of what was called the St. Louis Hegelian movement in philosophy, helping to popularize the ideas of the German philosopher in America. In 1867 Harris founded the *Journal of Speculative Philosophy* and commissioned immigrant Henry Brokmeyer, an intellectual comrade whom he had first met at the Mercantile Library, to translate some of Hegel's works for publication in its pages. Several important nineteenth-century American philosophers also made their debut in this journal. At least in certain intellectual circles, Anglo and German St. Louis had developed a fruitful symbiosis—a kind of synthesis that Hegel would have no doubt found gratifying.[47]

44. Troen, *Public and the Schools*, 65–78; "gerrymandering": ibid., 76. See also Olson, "St. Louis Germans," 96–108, although she exaggerates German indifference to language instruction.
45. King, *Great South*, 228; Troen, *Public and the Schools*, 64.
46. Troen, *Public and the Schools*, 99–110; Margot Ford McMillen, "Susan Elizabeth Blow," in Christensen et al., eds., *Dictionary of Missouri Biography*, 86–88.
47. Troen, *Public and the Schools*, 159–63; Denys P. Leighton, "William Torrey Harris," in Christensen et al., eds., *Dictionary of Missouri Biography*, 379–81.

South St. Louis Turnverein, Young Ladies' Class Bar Drill, c. 1910. Like the beer garden, the *turnverein*, or gymnastic society, was one of a number of conspicuous social and cultural institutions that St. Louis Germans brought with them to their adopted city. *Missouri Historical Society*

A similar synthesis was taking place on a much more prosaic level, as journalist Edward King observed in 1874: "At the more aristocratic and elegant of the German beer gardens, . . . many prominent American families may be seen on the concert evenings, drinking the amber fluid and listening to the music of Strauss, of Gungl, or Meyerbeer." King went on to say that Americans "no longer regard the custom as a dangerous German innovation," and that he found in St. Louis "many of the hearty features and graces of European life, which have been emphatically rejected by the native population of the more austere Eastern States." But this was not, he stressed, a one-way street: "[T]he German has borrowed many traits from his American fellow-citizen, and in another generation the fusion of races will be pretty thoroughly accomplished."[48]

At the time of Henry Shaw's death in 1889, St. Louis was still holding its own in the American urban hierarchy, though it later rose to fourth place only

48. King, *Great South*, 224.

because Brooklyn was swallowed by New York. But this was not a position that could be maintained for long. The leading sources of immigration were shifting from Germany and Ireland to southern and eastern Europe. St. Louis managed to attract a moderate number of Russian Jews and a fair number of Italians, but the Polish and other Slavic immigrants who formed the backbone of the labor force in heavy industry largely bypassed St. Louis. When the city experienced its shocking fall from fourth to sixth place in 1920, the two cities that had passed it, Detroit and Cleveland, boasted foreign populations in the 29–30 percent range, as did old rival Chicago. In St. Louis the immigrant proportion was not even half that large, only 13.4 percent. The same tendencies have continued throughout the twentieth century. By 1970, when Chicago was more than five times the size of St. Louis, immigrants and their American-born children together constituted barely 10 percent of the St. Louis population, a share far smaller than that represented by Chicago's 11-percent foreign-born population.[49] Of course, one question remains with respect to immigration and economic development: which is the chicken and which is the egg? I suspect that the causal arrows point in both directions, that mutually reinforcing tendencies are at work. Immigrants tend to go where the economy is most dynamic, but part of this dynamism comes from immigration itself. In that case, if the United States does indeed become a "majority-minority" society by the year 2050, there is only one thing that St. Louis has to fear: that current and future immigrants—especially the educated, enterprising twenty-first-century counterparts of Henry Shaw—will largely bypass what was once the Gateway City.

---

49. Primm, *Lion of the Valley,* 416–18; Holli and Jones, eds., *Biographical Dictionary of American Mayors,* 433, 439.

Part 2

Getting Ahead

Business, Science, Learning

The Economy of Nineteenth-Century St. Louis

James Neal Primm

In the seven decades of Henry Shaw's residence in St. Louis, the town (officially a city after its incorporation in 1823) grew in population from approximately 4,600 to more than 350,000—placing it, by 1880, sixth in rank in the United States. By 1900, St. Louis's 575,238 residents elevated it to fourth place among American cities—a distinction that, while reflective of the city's continued growth, nevertheless owed as much to the disappearance from the list of Brooklyn, which had since become incorporated into New York City.

Details of ranking aside, a simple question remains: What fueled nineteenth-century St. Louis's undeniably impressive expansion of wealth and population? For many at the time, geographic fortuity seemed explanation enough. St. Louis had taken full advantage of its location near the confluence of the Missouri and Mississippi rivers, with the mouth of the Illinois nearby. Its position at the highest point in the river system below which there were no rapids made it a natural transportation breakpoint. From the start, St. Louis was therefore ideally situated to command the western fur trade, and as settlers swarmed to the West in subsequent decades, the city grew to become a great commercial and distribution center. Though its commerce suffered cruelly during the Civil War and for a few years thereafter, St. Louis recovered in the 1870s and proceeded to become an industrial giant as well, until by 1900 it ranked fourth among American cities not only in population but also in industrial production.

Yet if location made this success story possible, it was people who made it happen. Women, limited in the explicit exercise of economic power by Missouri statutes, common law, and Victorian inhibitions, often pressed their influence in a relatively indirect manner, bringing family wealth and informed advice to energetic and talented husbands. By 1860 the efforts of a dozen or so such men, members of a new generation of civic leaders that followed the city's French and Creole founders, had made St. Louis a magnet for immigrants from the East, from the border states of Kentucky and Tennessee, and from Germany and

Ireland. These immigrants provided the city with a strong labor base, not dependent on slaves, and laid the basis for a growing and ambitious middle class. Outstate areas, meanwhile, were filling up with farmers and town-dwellers, and Missouri contained the largest free black population of the fifteen slave states.

While rivers had provided the city with its original transportation base, St. Louis's business and political leaders were, during this period of growth, acutely aware of the urgent need for rail connections to both the East and the West. They cajoled, bullied, and eventually persuaded local governments and private citizens—especially themselves—to invest heavily in statewide railroad construction; in the decade before the Civil War, the state spent more on rail construction than any other. Three of the four most important Missouri railroads terminated in St. Louis; the fourth was connected to the city by rail and water.

Despite St. Louis's remarkable economic expansion, another Midwestern city—Chicago—grew faster after midcentury. The comparison between the two great cities inevitably produced the legend that St. Louis had fallen behind through the indolence and wrongheadedness of its midcentury leadership. As this essay will show, such was not the case.

John A. Paxton, in his *St. Louis Directory and Register* of 1821, boasted that the town's merchants, especially in produce, furs, and peltries, carried on an "extensive trade, with the most distant part of the Republic."[1] As his counterparts did in subsequent St. Louis directories, Paxton puffed the town's glories, but he had walked the streets and interviewed the people. His compilation therefore presents a reasonable picture of St. Louis business just two years after Henry Shaw's arrival in the town.

Paxton's directory listed, among other enterprises, three newspapers, three large inns, thirteen shoemakers, twenty-eight carpenters, fifty-seven groceries (saloons), and twenty-seven attorneys. There were bookstores and binderies, gunsmiths, stonecutters, bricklayers, plasterers, bakers, cabinetmakers, wheelwrights, hatters, a brewery, a nail factory, a public market, professional musicians, and five billiard halls—the last a signature of the town's Creole heritage.[2]

Paxton enumerated 232 stone or brick houses (some doubling as businesses) and another 419 log or frame structures in the town. Through the efforts of Bishop Louis DuBourg, private subscribers (including many non-Catholics) had contributed funds to build a brick cathedral on the church lot, although the interior was not yet finished. Despite its incomplete state, Paxton described the

---

1. John A. Paxton, "Notes on St. Louis," in *The Early Histories of St. Louis,* ed. John Francis McDermott (St. Louis: St. Louis Historical Documents Foundation, 1952), 66.
2. Ibid., 67–68.

"Bird's-Eye View of St. Louis, Mo." (hand-colored lithograph by James T. Palmatary, 1858). While St. Louis's site—on elevated ground close to the meeting of the continent's two longest rivers—conferred great advantages, it was up to individuals to capitalize on those opportunities. By the time that James Palmatary drew his bird's-eye view of the newly expanded city, local entrepreneurs had turned river, road, and rail into a great system of trade and manufacture that brought together resources from the South and West with capital from the East. *Missouri Historical Society*

cathedral as a grand place, decorated with masterpieces painted by Renaissance artists.[3]

While Paxton's civic optimism moved him to estimate St. Louis's population at 5,500 (an increase over the 4,598 residents recorded in the just-completed decennial census), and while his tireless listing of individual structures and businesses suggested a prospering town, the St. Louis that Shaw saw as he looked around him in 1821 was in dismal economic straits. Hundreds of residents had died of malaria and recent business failures had removed many more.[4] The territorial Bank of Missouri, despite its favored status as a U.S. Treasury depository, had lent heavily on real estate, and the collapse of property values had led to its failure. In August alone, the sheriff had sold 105 lots in St. Louis and 14,000 acres in the county for nonpayment of taxes. (John Mullanphy, who had made a fortune at the end of the War of 1812 by cornering the New Orleans cotton market, acquired thousands of acres in north St. Louis County at these and later

---

3. Ibid., 65. If such paintings had ever hung in the unfinished cathedral, they left with Bishop DuBourg, who moved to New Orleans in 1822.
4. See Dorothy B. Dorsey, "The Panic of 1819 in Missouri," *Missouri Historical Review* 29 (January 1935). See also James Neal Primm, *Economic Policy in the Development of a Western State: Missouri, 1820–1860* (Cambridge: Harvard University Press, 1954), 1–3.

tax sales.)[5] Meanwhile, notes issued by Missouri's "Loan Office," established in the previous year by the General Assembly, were now worthless outside the state, while local merchants either refused them or discounted them heavily.[6] By 1823, the year of St. Louis's incorporation, property values within the city limits had fallen to $810,000 (from $1.21 million in 1818) and the city's population had shrunk to about 3,000.

Economic recovery was well under way by 1824 and migration to Missouri resumed. Thousands of families passed through St. Louis each year, headed for the rich bottom lands of central and western Missouri and the Salt River valley northwest of St. Louis. Many of these pioneer farmers, chiefly from Kentucky and Tennessee, traveled light and purchased their "necessaries" in St. Louis before moving on.

The change was nowhere more noticeable than on the city's levee. Steamboating had had a modest beginning in 1817, when the little *Zebulon Pike* limped into the St. Louis harbor, its engine too weak to fight the current without the assistance of pole men lined up along each side of its deck. By the mid-1820s steamboats were navigating the Mississippi, Missouri, and Illinois rivers, carrying merchandise from St. Louis to frontier towns such as Rocheport, Lexington, and Independence on the Missouri; Hannibal on the upper Mississippi; and Beardstown on the Illinois. The boats returned to St. Louis loaded with produce, some of it for local consumption, but most of it destined for the Ohio River ports or New Orleans. As a transportation breakpoint, St. Louis prospered because larger vessels had difficulty navigating either the turbulent Missouri or the dangerous rapids and frequent low water of the upper Mississippi. Using the smaller shallow-draft upriver boats on the lower river would have been uneconomic. Transshipments at St. Louis created a demand for labor and services at the wharves and for commission merchants and forwarding houses to move the goods and produce to their market.[7]

In 1826 at least one steamboat a day entered the St. Louis harbor during the ice-free months, and in 1827, 290 steamboats anchored there. Bernard Pratte, a fur trader, sent the *Yellowstone* to the upper Missouri in 1832. Keelboats were

---

5. John Ray Cable, *The Bank of the State of Missouri* (New York: Columbia University Press, 1923), 67–72.

6. Primm, *Economic Policy*, 13–17, 131. The Loan Office Act was declared unconstitutional by the Missouri courts in 1822 and by the U.S. Supreme Court, in the case of *Craig* v. *Missouri*, in 1828.

7. See George R. Taylor, *The Transportation Revolution* (New York: Rinehart, 1951), 166. For a fuller discussion, see Louis C. Hunter, *Steamboats on the Western River: An Economic and Technological History* (Cambridge: Harvard University Press, 1949), and James Neal Primm, *Lion of the Valley, St. Louis, Missouri, 1764–1980*, 3d ed. (St. Louis: Missouri Historical Society Press, 1998), 134–35.

still laboring up the rivers, but steamboats reduced downstream time to New Orleans by two-thirds, and upstream time by seven-eighths in 1830. By 1836, 1,355 steamboats arrived, and by 1841 St. Louis had overtaken the Ohio River ports. With 186 steamboats (nearly half of them owned by local merchants and captains) making 1,928 landings and delivering 262,681 tons of goods, St. Louis stood second only to New Orleans in annual volume of traffic.[8]

The first of the great commercial enterprises that helped to spur this early river traffic was the fur trade, which had given birth to St. Louis and which continued to dominate the city's economy three-quarters of a century later. Like other businesses, fur trading suffered from low demand during the panic and depression (1819–1823), and these losses were compounded when the Arikara Indians closed the upper Missouri after the death of their friend Manuel Lisa in 1820. Farther upstream, the Blackfeet, who had hated the Americans since Lewis and Clark had passed their way, represented another impediment. Even Lisa had been unable to pacify them. But fur prices were recovering by 1823, and William H. Ashley, after his party was turned back by the Arikaras, pioneered an overland route to the Central Rockies, following the Platte and the North Platte directly to the beaver bonanza.[9]

As profits grew, the fur trade began to take on the character of a regularized (if still cutthroat) corporate business. In 1826 Ashley sold his firm, later known as the Rocky Mountain Fur Company, to William Sublette, David Jackson, and Jedediah Smith. Ashley continued to finance expeditions, often very profitably. A year later Bernard Pratte and Pierre Chouteau Jr. sold their own concern to John Jacob Astor's American Fur Company. Astor then bought Kenneth McKenzie's Columbia Fur Company, another St. Louis firm, and combined the two under McKenzie as the "Upper Missouri Outfit" of the American Fur Company. McKenzie, a legendary figure known as "the King of the Upper Missouri," had cultivated good relations with the Blackfeet, and was able to trade (but not trap) in their territory.[10] In the early 1830s the Outfit, the Rocky Mountain Fur Company, and others contended fiercely for the mountain trade. With demand for beaver diminishing, Pratte and Chouteau then consolidated the American Fur Company's Western Department and the Rocky Mountain Fur Company under their ownership in 1834. Chouteau purchased Pratte's interest in 1838, the latter having shifted his interest to St. Louis real estate and

---

8. Primm, *Lion of the Valley*, 134–35.
9. David Lavender, *The American Heritage History of the Great West* (New York: Simon & Schuster, 1965), 174–76; William B. Nester, *From Mountain Man to Millionaire: The Bold and Dashing Life of Robert Campbell* (Columbia: University of Missouri Press, 1999), 13–15.
10. McKenzie, a ruthless competitor, was the nephew of the Canadian explorer Alexander Mackenzie, who reached the Pacific by an overland route before Lewis and Clark.

politics. With his firm still known informally as the American Fur Company, Chouteau dominated the buffalo robe trade from posts on the upper Missouri and in the Laramie Mountains.[11]

Pratte, Chouteau, Ashley, Sublette, Robert Campbell, and others prospered amidst these corporate maneuvers. Ashley's sale, for instance, had netted him a reported $50,000.[12] They risked losing everything to accident or Indians on every expedition, but they kept their scalps, while many of their trappers did not.

Such profits from remote trading operations exerted a significant influence on St. Louis. At times the fur trade yielded more than 100 percent in profits in a year, although the average return was between 30 and 40 percent. According to one estimate, the fur trade had brought $10 million into the Missouri economy by 1847. Its leaders often served their community as well, during and after their years in the mountains. Sublette, Campbell, and McKenzie operated outfitting stores for years. Pratte, Chouteau, Ashley, and Campbell were active in banking. Campbell, who in 1832 had carried a wounded Sublette on his back during a running gun battle with the Gros Ventres at the Pierre's Hole rendezvous, became president of the Bank of the State of Missouri in 1846 and later bought and operated the swank Southern Hotel in St. Louis.[13]

Expanding trade territories added to St. Louis's renewed prosperity, as well. To the southwest, the newly independent Mexico opened its northern door to Missouri traders in 1821. Even as the eastern terminus of the Santa Fe Trail moved from Franklin, in central Missouri, to Independence and finally to Westport, on the state's western border, St. Louis remained an engine of the southwest trade. Some of the larger expeditions were organized in the city, its merchants invested in the western towns, its wholesalers supplied the western outfitters, and the Murphy wagons that carried the goods were made in St. Louis. More important, because the terms of trade favored the Americans, the Mexicans had to augment their shipments of furs and mules with silver, most of which eventually came to St. Louis. In 1829 the trade furnished $200,000 in silver to the city's economy; ten years later, the same amount arrived in five weeks. Western merchants and Mexicans from Santa Fe and Chihuahua often maintained balances at St. Louis's private banks.[14]

Banking policy, which determined the actual circulation of money in and around the city, was as important a contributor to the cycles of growth and

11. Nester, *From Mountain Man to Millionaire*, 82–105.
12. Ibid., 20–21. As lieutenant governor of Missouri, Ashley's salary had been one thousand dollars a year.
13. Ibid., 166, 172; Timothy W. Hubbard and Lewis E. Davids, *Banking in Mid-America: A History of Missouri's Banks* (Washington, D.C.: Public Affairs Press, 1969), 58.
14. Hubbard and Davids, *Banking in Mid-America*, 12, 63–64.

contraction that affected antebellum St. Louis as were raw commodities like fur. A brief look at the difficulties encountered by the city's financial institutions during the 1830s will demonstrate their central role in the city's economy.

The Bank of the United States established a branch at St. Louis in 1829, providing a sound currency and funds for business operations and expansion. John O'Fallon, nephew of territorial governor William Clark and a major supplier for the U.S. Army, was named president of the branch, and William H. Ashley, Thomas Biddle (brother of Nicholas Biddle, the president of the parent bank), Pierre Chouteau Jr., James Clemens, Peter Lindell, and John Mullanphy were among its prestigious directors. By 1832 the branch's discounts (loans) had reached $883,000, and its notes and the inflow of Sonoran silver supported business transactions, employment, and property values.[15]

Unfortunately, after President Jackson vetoed the re-charter of the Bank of the United States in 1832, the St. Louis branch—following Philadelphia's lead—reduced its discounts, curtailing the money supply. In 1835 the branch closed its doors. The Cincinnati Commercial Agency, encouraged by Senator Thomas Hart Benton and John O'Fallon, then opened a branch in St. Louis. As the branch bank had done, the Agency encouraged trade at St. Louis by accepting the notes of the Bank of Illinois, often tendered by Illinois retailers to St. Louis wholesalers for farm equipment and other manufactured goods.[16]

In 1835, however, the state of Illinois went on a canal and railroad construction spending bender, and its state bank lent heavily to these projects, flooding the state with banknotes which it could not redeem in specie. As these notes and others of their kind poured into St. Louis, pressure built for chartering a state bank in Missouri. Hard-money disciples of Senator Benton, once opposed to all banks of issue, began to consider a conservative, specie-paying bank preferable to the "rag-money" they were having to digest. In 1837 the General Assembly chartered the Bank of the State of Missouri, with headquarters in St. Louis and branches initially at Fayette and Lexington. The bank bill had been managed by Democrats, and the charter mollified the "hards" by putting Benton's picture on the banknotes, reserving half of the bank's stock for state ownership, limiting the circulation to denominations of $10 or more, and empowering the legislature to elect the president and half of the directors.[17] John B. Smith of St. Louis was chosen its president, with O'Fallon, Sublette, and Edward Walsh among the directors. O'Fallon's selection especially was reassuring, given his masterful management of the Branch Bank of the United States.[18]

15. Cable, *Bank of the State of Missouri*, 85–95.
16. Ibid., 95–99.
17. Ibid., 122–44; Hubbard and Davids, *Banking in Mid-America*, 51–63.
18. Cable, *Bank of the State of Missouri*, 161.

Bank of the State of Missouri to Bank of America. $20. November 1, 1838 (banknote engraved by Draper, Toppan, Longacre and Co.). A wave of financial panics—including the depression of 1837—led to increasing regulation of the city's banks and lending institutions. The Bank of the State of Missouri was established to rein in the supply of cheap money that had contributed to this latest financial crisis. *Missouri Historical Society*

Smith and his directors set to work. First, the state bank bought the Cincinnati Agency's buildings and began to replace its notes with its own. On November 21, 1837, its assets totaled $1,917,978, including discounts of $738,845 and $508,696 in deposits. It had already contracted its discounts, and it continued to do so, increasing the share of out-of-state banks in St. Louis's circulation. In November 1839 it struck a chilling blow, announcing that it would no longer accept the notes of nonspecie-paying banks. This further depreciated the already-discounted out-of-state notes, some of which would no longer pass at all.[19]

St. Louis merchants, led by Chouteau, declared war on the bank. In their view, it had placed its own survival ahead of the merchants' credit needs. In the midst of bank failures everywhere, the Bank of the State of Missouri was hailed as a Rock of Gibraltar. But to its critics, Gibraltar had stifled economic growth by failing to provide a circulating medium. Its note circulation shrank from $1.75 million to $352,000 in 1840. Merchants and politicians fought back with a variety of panaceas, including an ill-fated issuance of city scrip in 1841, which had an even shorter life than the old Loan Office certificates. Chouteau, George Collier, and nine other businessmen offered to indemnify the bank for any losses incurred by accepting Illinois banknotes, but the bank rejected the offer.[20]

---

19. Ibid., 179–87.
20. Ibid., 180; Hubbard and Davids, *Banking in Mid-America*, 60.

In 1840 James Lucas's St. Louis Gaslight Company and several insurance companies stepped into the breach by accepting notes of suspended banks on deposit, and borrowing, lending, and spending them, thereby assuming the functions of banks. The "hards" in the Missouri legislature responded with bills repealing the charters of companies that violated their charters and providing that "no corporation, money broker, or exchange dealer should receive any banknotes of less than $10 denomination."[21]

At the height of the crisis, Jesse Baker, one of Henry Shaw's business associates, wrote to Shaw at Naples, Italy, that there would be little business done for some time: "[T]he fact is there is no money . . . the currency here is in such wretched condition . . . our circulation here is mostly Illinois and Indiana money but I can see but little of either."[22] The currency problem for merchants was incredibly complicated, as the accounts of one St. Louis businessman, Hudson Bridge, will illustrate. On March 2, 1842, J. B. Evans of Cincinnati wrote to Bridge that the "Indiana $5 bill you sent us is worth 37 1/2 cents on the dollar," and that "One $3 bill on the Marietta Bank is not very good." On December 20, 1842, Jewett and Hitchcock of Springfield, Illinois, sent Bridge a cash payment consisting, they wrote, of the following notes: "Bank of Missouri, $250; Bank of Indiana, $257; Ohio Life Ins. and Trust, $15; Bank of Kentucky, $30; Ohio bank, $13; South Carolina banks, $17; Virginia banks, $15; Wisconsin Ins. Co., $32; St. Louis City Scrip, $5; Maine banks, $20; Louisville Bank, $20."[23]

Private bankers, or note brokers, who operated "without benefit of law" as one authority put it, were more important than the erring gaslight and insurance companies.[24] Louis Benoist, George Budd, and others discounted out-of-state notes and lent them at interest on short-term commercial paper. Since they were not chartered, these private businesses were not creatures of the state and not inhibited from discounting banknotes. Since the brokers would not make real estate loans, certain wealthy merchants such as Shaw, Peter Lindell, and eventually Hudson Bridge provided some relief by lending against real property. Usually it was difficult to tell whether they preferred repayment or forfeiture.[25] By 1842 the Bank of Illinois was about to close, with its notes at a 45 percent discount. Benoist was swamped by this paper, and had to close temporarily. The discounted notes had deferred but not prevented a substantial liquidation,

---

21. Cable, *Bank of the State of Missouri*, 185; Primm, *Economic Policy*, 47.
22. Jesse Baker to Henry Shaw, March 22, 1841, Henry Shaw Papers, Missouri Botanical Garden Archives, St. Louis.
23. Hudson Bridge Papers, Missouri Historical Society Archives, St. Louis.
24. Hubbard and Davids, *Banking in Mid-America*, 63.
25. Primm, *Lion of the Valley*, 197. See also Louis V. Bogy to Henry Shaw, October 5, 1842, Shaw Papers.

and dozens of shops and stores closed. Shaw's St. Louis correspondents while he was abroad in 1841 and 1842 usually had one or more failures to report.

The distress was soon ameliorated in St. Louis, in part by Mexican silver and in part by an approaching immigrant tide. Irish, Germans, and a host of energetic young men from the border and eastern states and the British Isles entered the St. Louis economy in the late 1830s and early 1840s. Among these were such future titans as Wayman Crow and Derrick A. January from Kentucky; James Yeatman from Tennessee; Thomas Allen, Oliver and Giles Filley, Hudson Bridge, and Carlos Greeley from New England; and Gerard B. Allen from Ireland.[26]

Economic recovery was well under way by 1843, nationally and in the West. Between 1843 and 1847 exports of flour from St. Louis quadrupled, and shipments of buffalo hides, salt pork, and hemp bales nearly doubled. Hide hunters were slaughtering the herds of the northern plains to deliver buffalo hides to Pierre Chouteau at Fort Laramie and Fort Pierre, and some forty-eight hundred planters in western Missouri were growing nearly one-third of the hemp produced in the nation. Receipts of wheat at St. Louis more than tripled between 1842 and 1846, to more than 1.8 million bushels. The grain was delivered by steamboats from the Illinois River valley, the Salt River area of northeast Missouri, and from Boonslick and western Missouri. St. Louis County itself grew 98,439 bushels in 1850, fourth among Missouri counties. As was true of the other nearby counties, this wheat was delivered to St. Louis mills by wagon. Missouri and Ohio were the national leaders in wheat production at midcentury, yet because of the rigorous quality control enforced by the St. Louis Miller's Exchange and the lower moisture content of western wheat, St. Louis flour commanded a premium of fifty cents a barrel over Cincinnati's. The Miller's Exchange, the first grain exchange in the nation, merged with the Chamber of Commerce in 1850 to form the Merchant's Exchange.[27]

Daniel Page's Star Mill produced 70,000 barrels of flour in 1847. Fourteen other mills brought St. Louis's total to 365,000 barrels. To move the produce and goods, there were fifty commission houses in St. Louis, in addition to speculative commodity dealers such as Peter Lindell and Henry Shaw, who bought tobacco, flax, beeswax, and other products from country dealers for sale in British and East Coast markets via New Orleans.[28]

---

26. Primm, *Lion of the Valley*, 225–26.
27. Ibid., 193. See also Walter B. Stevens, *St. Louis, the Fourth City, 1764–1909* (St. Louis: S. J. Clarke, 1909), 1:681–82.
28. See J. Hoole to Henry Shaw, September 26, 1842, and William Hempstead to Henry Shaw, December 31, 1842, Shaw Papers. There are dozens of letters in the Shaw Papers illustrating the point.

During the 1840s St. Louis's population nearly quintupled, from 16,439 to 77,680, making it the nation's eighth largest city in 1850, after New York, Philadelphia, Baltimore, Boston, New Orleans, Cincinnati, and Brooklyn.[29] Germans, who had arrived by the thousands in the 1830s, settling mostly in the Missouri River valley west of St. Louis, continued to come in the 1840s, especially after 1848. Poverty and even the threat of starvation drove young men and women out of Ireland in the 1840s, nearly ten thousand of them to St. Louis. The ratio of men to women, three to two in 1840, declined only slightly during the decade, despite the balancing effect of a large natural increase. Among the city's slave population, on the other hand, a majority were females—a result of the high marketability of young males for the rigorous, labor-intensive Louisiana sugarcane and Missouri hemp plantations.[30]

Tied in with all this growth was St. Louis's continued role as a center of western transportation. The rivers, of course, remained key to that role. In 1857 there were 3,443 landings by boats averaging 500 to 600 tons in cargo capacity. The volume of goods delivered had tripled since 1844. By comparison, the city's chief rivals, New Orleans and Cincinnati, had 2,745 and 2,703 landings, respectively, in 1857. St. Louis also led in steamboat ownership and thirty of its machine shops were manufacturing boat machinery. While in the early years steamboats were usually owned by wholesale merchants, by the 1850s their captains or investors owned most of the boats.[31]

The boats came from all directions. Between 1847 and 1854, the upper Mississippi towns of Hannibal, Keokuk, Davenport, Rock Island, Galena, Dubuque, and others accounted for 24.5 percent of the arrivals at St. Louis; 22.8 percent came from Beardstown and Pekin on the Illinois River; 16.2 percent from Ohio River ports (chiefly Cincinnati and Louisville); 12.7 percent from New Orleans; and 11.4 percent from Missouri River ports such as Rocheport, Lexington, Independence, Fort Leavenworth, and the northern fur-and-hide boats. The remainder came from the lower Mississippi above New Orleans.[32]

Just as the steamboat era peaked, technology was planning its destruction. Boats were already towing huge rafts of logs from Minnesota to Hannibal and St. Louis in the 1850s; it was not long before someone would discover that only

---

29. Joseph C. G. Kennedy, Superintendent, *Preliminary Report on the 8th Census, Washington, 1860*, 342–45.
30. George Kellner, "The German Element on the Urban Frontier: St. Louis, 1830–1860" (Ph.D. diss., University of Missouri, Columbia, 1973); Primm, *Lion of the Valley,* 179–80.
31. Primm, *Lion of the Valley,* 160–64.
32. *Western Journal of Agriculture, Manufacture, and Mechanic Arts* (St. Louis: M. Tarver & T. F. Risk, 1848), 1:55; Primm, *Lion of the Valley,* 161. These data can be misleading if it is not understood that the tonnage of the lower riverboats was much larger than upriver craft.

the engines were needed. Barges would do for bulky, low-unit-cost items. Yet it was another transportation innovation, the iron horse, that exercised the most devastating impact on river commerce and ushered in a new era of St. Louis commerce.

That the impact of the railroad was not felt even sooner than it was in St. Louis says more about financial timing than it does about technological readiness. St. Louisans first contracted railroad fever in 1836, only a few years after railroads were introduced in the United States. Abel R. Corbin, editor of the *St. Louis Missouri Argus,* argued that the city's future as the trade center of the Mississippi valley depended on state-aided railroad construction. On March 11, he proposed that five railroads be built from St. Louis: one to a source of wood or coal; a second to the iron and lead deposits south and west of the city; a third to Fayette in central Missouri; a fourth to southwest Missouri; and a fifth through Franklin County to Independence.[33]

After state-aided railroad construction had been endorsed at a railroad convention in St. Louis and by most of the state's newspapers, the General Assembly in January 1837 chartered eighteen railroads, the most important of which ran from St. Louis to Potosi and its lead mines and to Caledonia near Iron Mountain and Pilot Knob. Others ran from Hannibal to St. Joseph and from Boone County to the Mississippi. Short-line roads made up the rest. Counties were authorized to tax real estate to support construction, but the state made no financial commitment; it had neither the capital nor any means of getting it. State revenue for the previous biennium had been less than $200,000. Three more railroads were chartered in 1839, but as the economy deteriorated during the depression, the champions of railroad construction changed their tune. The *Missouri Argus* congratulated the state in November 1840 for not building railroads, noting that Illinois had created a debt of more than $13 million with only a half-finished canal, a statehouse, and a twenty-four-mile-long railroad to show for it.[34]

With the mid-1840s economic recovery, the passion for railroads returned. In 1847 the Missouri legislature chartered the Hannibal and St. Joseph Railroad, and a year later it authorized St. Louis to invest in the Ohio and Mississippi, which would connect the city to Cincinnati and thus to Baltimore via the Baltimore and Ohio. Then, in 1849, Missouri's Thomas Hart Benton introduced a bill in the U.S. Senate proposing the most ambitious line of all: a railroad from St. Louis to San Francisco, to be built and owned by the federal government and financed by the sale of public lands. A few weeks later the Missouri legislature

33. Primm, *Economic Policy,* 77–78.
34. Ibid., 79–87.

chartered the Pacific Railroad, to be constructed from St. Louis through Jefferson City to the state's western border, with California the ultimate objective.[35]

Thomas Allen, a commercial real estate investor, was the engineer of the Pacific charter. Allen, a Massachusetts native, was an attorney and a gifted writer and publisher who had held the government printing contract in Washington until the political winds changed. In 1842 he opened a law office in St. Louis, but before he had clients he met and married Ann Russell, whose father, William, had made a fortune in Missouri and Arkansas real estate. The marriage altered his career plans and focused his attention on St. Louis's future.[36]

Memphis and New Orleans were also in the transcontinental railroad hunt, and northeastern interests were pushing for one with a Great Lakes terminus, which would deliver the heralded Asian trade to their own front door. This would give wings to upstart Chicago, which had not yet dented St. Louis's position in the western trade. In the war between the cities conducted in the press, the St. Louis papers reserved their sniping for Cincinnati and New Orleans, with an occasional warning shot for Louisville and Pittsburgh. Chicago had held a railroad convention in 1847, and now, with barely a third of St. Louis's population and business, it entered the arena. Memphis was not much larger than Chicago, but it had strong mid-South support. It was time for St. Louis to move.[37]

Allen prepared and circulated nationally a brochure lauding St. Louis as a transportation and trading center, and Mayor James Barry invited all the states

---

35. Ibid., 93–97; Ethel Osborne, "Missouri's Interest in the Transcontinental Railroad Movement, 1849–1855" (M.A. thesis, University of Missouri, 1928), 6–23.

36. Primm, *Lion of the Valley*, 206; Richard Edwards and M. Hopewell, *Edwards's Great West and Her Commercial Metropolis* (St. Louis: Edwards's Monthly, 1860), 437–39. Ann Russell Allen was just one of several wealthy nineteenth-century St. Louis women who exerted a significant but underreported influence on the growing city. Unlike Marie Thérèse Chouteau, who had bought and sold commodities on her account under Spanish rule, mid-Victorian-era St. Louis women were denied significant economic roles by statutory and common law. Because they were unable to transact business except through husbands or male relatives, their names do not appear in the hundreds of St. Louis entries in the *R. G. Dun Credit Reports* from 1840 to 1880. Yet some of the city's largest landowners—Allen, John O'Fallon, U. S. Harney, and Richard Graham—built their fortunes through marriage to women of means, the last two to John Mullanphy's daughters. Anne Hunt, who had inherited the late Charles Lucas's Normandy property from her father, was herself one of the largest landowners in St. Louis County, but earned no separate listing on the tax rolls. Shaw's sister, Caroline, who ably conducted her brother's extensive commodities trading business during his long sojourns abroad, depended on his attorney or his friend, Peter Lindell, to endorse her transactions. Legal convention, too, has hidden the full historical role of St. Louis women; to cite but one example, Irene Emerson, the actual owner of the slave Dred Scott, was replaced by her brother as defendant in the case of *Dred Scott v. Sandford*. For more on the city's notable women, see Katharine T. Corbett, *In Her Place: A Guide to St. Louis Women's History* (St. Louis: Missouri Historical Society Press, 1999).

37. Paul W. Gates, "The Railroads of Missouri, 1850–1870, *Missouri Historical Review* 26 (January 1932): 128–29.

to send delegates to the city's own railroad convention in October 1849. Delegates from fifteen states attended, 1,056 in all, but four-fifths of them were from Missouri and Illinois, while Indiana and Iowa furnished most of the rest. Here, Senator Benton delivered his famous "there is the East, there is India" speech, advocating a railroad from St. Louis to San Francisco, with a branch to the Columbia River. Allen wrote the convention's memorial to Congress, urging it to build along the route recommended to Benton by the senator's son-in-law, Lieutenant Colonel John C. Fremont, who had explored the central Sierra and Rocky Mountain passes for Benton and some St. Louis businessmen. This route crossed the Sierras above the Sacramento River valley and then ran eastward past the headwaters of the Rio Grande and the upper Arkansas River valley and Kansas River valley to St. Louis. The entire route lay between the 38th and 39th parallels.[38]

On January 21, 1850, the incorporators of the Pacific Railroad met to seek subscribers to its common stock, so that officers could be elected. James Lucas offered to be one of three to pledge a total of $100,000 and John O'Fallon and Daniel Page joined him. Thomas Allen, George Collier, James Harrison, and James Yeatman made large pledges, and after a twelve-day door-to-door campaign the promoters had $350,000 in hand. Soon thereafter, Mayor Luther M. Kennett and the council committed the city to $500,000 in Pacific bonds.[39]

Allen then stumped the state, meeting with businessmen and politicians in the county-seat towns, not only to sell Pacific stock, but to create sentiment for a comprehensive system of railroads in Missouri, to be supported by public as well as private investment. Stockholders were not promised large returns; instead, their rewards would come from accessible markets, general prosperity, and a larger tax base. In the fall, Allen was elected to the state senate, avowedly to craft and manage railroad bills. As chairman of the Internal Improvements Committee, he proposed a system of six state-aided railroads, three of them with St. Louis terminals and the others to connect with the first three and thence to the Missouri or Mississippi.[40]

In February 1851 Allen's railroad bills were approved, providing state guarantees for $2 million in Pacific and $1.5 million in Hannibal and St. Joseph

---

38. Robert E. Riegel, "The Missouri Pacific Railroad to 1879," *Missouri Historical Review* 18 (October 1923): 4–5; Osborne, "Missouri's Interest," 32–36. Senator Stephen A. Douglas of Illinois, determined to deny the prize to St. Louis, favored a route from Council Bluffs, Iowa, through South Pass to the Pacific. Branch lines would connect the main line to Chicago, St. Louis, and Memphis. The Fremont route was straighter and shorter, with equivalent gradients, but Douglas's warning that Congress would not support a single-city eastern terminus proved prophetic.
39. Primm, *Lion of the Valley,* 205–6.
40. Ibid., 206–7.

railroad bonds. Allen then traveled to Washington, where Congress was considering a land grant for the Illinois Central Railroad. The Illinois promoters won the first such grant ever made, but Allen and the Pacific were turned down. He did learn how it was done, however; he hired two veteran lobbyists for $15,000, and soon Congress awarded each Missouri railroad alternate sections of land six miles deep on each side of its right-of-way. The roads were to sell this land to pay for construction and start-up costs. The Hannibal and St. Joseph would intersect the North Missouri, which would lay track from St. Louis to the Iowa line. As Allen and O'Fallon saw it, the Hannibal and St. Joseph would feed northwestern commodities to St. Louis via the North Missouri or the Mississippi River from Hannibal.[41]

By 1860 Missouri had committed $7 million in bond guarantees to the Pacific, $4.35 million to the North Missouri, $3 million to the Hannibal and St. Joseph, $3.5 million to the St. Louis and Iron Mountain, and $3.9 million to the Southwest Branch of the Pacific. The last would run from the Pacific Railroad at Franklin (now Pacific) through Rolla to Springfield. It had been added to the mix to win support for the system from that part of the state. It also won a federal land grant. Smaller bond guarantees were awarded to two short-line companies, one in southeast Missouri and one in Platte County, just north of the great bend of the Missouri River. In a decade, Missouri had pledged its credit for these railroads in the amount of $23.1 million. In addition, it was responsible for the interest on the bonds if the railroads defaulted. By 1860, that was precisely what happened. All of the railroads were in default except the Hannibal and St. Joseph, which was financially stronger because of its valuable land grant and which lay along a route more attractive to eastern investors seeking a ready connection to the Northeast's commercial orbit.[42]

Missouri, the eighth largest state in population in 1860, had the fourth largest debt. Only Virginia among the thirty-three states had amassed a larger debt since 1845. St. Louis city and county had invested $6.15 million in Missouri railroads and $500,000 in the Ohio and Mississippi. Individuals in the city had invested $1.38 million in the Pacific, $500,000 in the Ohio and Mississippi, $245,000 in the Iron Mountain, and lesser amounts in the other roads. Private banks had made substantial loans, especially to the Ohio and Mississippi.[43]

The business leadership, mobilized by Thomas Allen, devoted itself to what it believed to be an indispensable civic enterprise. John O'Fallon, James Harrison,

41. *Laws of the State of Missouri, 1850* (City of Jefferson, James Lusk, Public Printer, 1851), 265–68.
42. John W. Million, *State Aid to Railways in Missouri* (Chicago: University of Chicago Press, 1896), 91.
43. Benjamin U. Ratchford, *American State Debts* (Durham: Duke University Press, 1941), 124–27; Million, *State Aid to Railways*, 242–43.

PACIFIC RAILROAD.

The most Reliable and Direct Route for
KANSAS CITY, LEAVENWORTH & ST. JOSEPH.
Trains Leave St. Louis as follows:

MAIL TRAIN—Daily, at 9:00 A. M., stopping at all Stations and running through to Syracuse.
EXPRESS TRAIN—Daily (except Sunday) at 3:50 P. M., stopping at principal stations, and running to Jefferson City only.
FRANKLIN ACCOMMODATION—Every day (except Sunday) at 5¼ P. M.

☞Through to Jefferson City in SIX HOURS.

The trains of this road connect at Jefferson City with a daily line of first class

PASSENGER PACKETS,

Which leave immediately on arrival of Express Train for all points on the Missouri River, connecting at Kansas City and Leavenworth with daily lines of stages for Fort Riley and the interior.

Passengers taking this line will avoid detention at St. Louis, save 50 miles in distance, and escape the delays of 155 miles of difficult river navigation, and make the trip from

ST. LOUIS to KANSAS CITY in 48 Hours!

From Syracuse, stages leave daily at 7:30 P. M., (on arrival of mail train from St. Louis), for Springfield (South-West Missouri), Independence (through Georgetown and Warrensburg), and on every Monday and Thursday P. M. the Overland mail Stages of Butterfield & Co.

Through to San Francisco in twenty-three days!

Fare as Low as by any other Route.

Passengers arriving on Eastern and Northern (Morning) trains, have plenty of time to connect with the Express Train of this road.

Passengers arriving at Jefferson City, pass directly aboard the packets, and proceed on their route without incurring extra expense.

Baggage checked to its destination, and transferred to boat free of charge.

Through Tickets may be obtained, securing meals and berths on boat, at the Passenger Depot, corner of Seventh and Poplar streets, or at the Company's Through Ticket Office, No. 42 Fourth street, under the Planters' House; also, at all the principal Railroad Offices in the United States and Canadas.

E. W. WALLACE, Gen. Ticket Agent.
T. McKISSOCK, Superintendent.

Advertisement for the Pacific Railroad (from "Edwards Programme and History of the Fair," 1859). With the political support of Thomas Hart Benton and financial backing from Thomas Allen and others, the Pacific Railroad was the centerpiece of Missouri's aggressive bid to become a major railroad state in the 1850s. While Chicago would ultimately capture a larger region with its own lines, Missouri's tracks helped to ensure St. Louis's continuing domination of the southwestern trade territory. *Missouri Historical Society*

Jules Vallé, Hudson Bridge, and James Lucas organized the St. Louis and Iron Mountain in 1851. Harrison and Vallé had good reason to expedite the flow of pigs and blooms from their furnaces at Iron Mountain to their St. Louis mill, and so did Bridge, whose Empire Stove Works needed a steady flow of raw material. Bridge imported iron from Jevons and Company in Liverpool; he owned an iron bank in southern Kentucky and at times his agents purchased stoves in Albany and upstate New York for resale. The last was often difficult, because the New Yorkers did not like to sell to St. Louis dealers, with whom they were in direct competition on the upper Mississippi.[44]

St. Charles merchants organized the North Missouri Railroad, but it was quickly taken over by O'Fallon, who became its first president in 1853. On the east side, the directorate of the Ohio and Mississippi western branch (the St. Louis and Vincennes) included St. Louisans Daniel Page, Andrew Christy, Samuel Gaty, and Charles P. Chouteau. The indefatigable O'Fallon was president. Christy was co-owner of the Wiggins Ferry, which transferred cargoes to and from the railroad terminal and the St. Louis wharves. Page invested personally, and his Page and Bacon bank financed the construction of the western branch. The Ohio and Mississippi reached Illinoistown (East St. Louis) in 1857, giving St. Louis new markets in southern Illinois and Indiana as well as its East Coast connection.[45]

Railroad investors and officers often, though not always, linked their expectation of economic gain with a sense of civic virtue. Contributing to the city's progress (and being recognized for it) was a powerful motive. Colonel John O'Fallon, a veteran of Tippecanoe and the War of 1812, had acquired large real estate holdings by marriage in 1821, and the growth of the city and real estate appreciation, augmented by his profits as an army supplier, had made him rich. He had already given large amounts to charity by 1851, and by 1860 his gifts to the city and to various charities exceeded $1 million, including $100,000 for the founding of O'Fallon Polytechnic Institute.[46] James Lucas—who was far wealthier than O'Fallon—Thomas Allen, Daniel Page, Hudson Bridge, and other railroad promoters and investors may have been expectant capitalists, but they did not expect to gain directly from their investments of time and money for years, if ever.

Despite delays caused by wars between the Corkers and Kerrymen of the construction crews, difficult terrain, and a bridge collapse over the Gasconade River, the Pacific reached the state capital by 1855. But construction slowed, especially after the Panic of 1857 hammered the bond market. The Pacific's land

---

44. James Harrison to Hudson Bridge, June 8, 1844, Bridge Papers.
45. Primm, *Lion of the Valley*, 219–33.
46. Edwards and Hopewell, *Edwards's Great West*, 79–82.

grant was worth about a third of that of the Hannibal and St. Joseph's, because some of its land was inferior and located far from the right of way. In 1861 the tracks were still some eighty miles from the Kansas border, and Confederate raiders and guerrillas helped keep them there until 1865. The Hannibal and St. Joseph, having easier gradients, better soil, and more money, was completed in 1859, making St. Joseph for a golden moment the westernmost point in the United States served by railroad and telegraph. To the south, the St. Louis and Iron Mountain connected St. Louis to the largest known iron ore deposits in the nation, and the Southwest Branch had reached Rolla, a short wagon-haul from the Meramec Iron Works, which had supplied St. Louis foundries for three decades by horse-and-mule-killing wagon hauls. The North Missouri had reached the Hannibal and St. Joseph tracks at Macon, but its freight and passengers still had to cross the Missouri River at St. Charles by ferry.[47]

Altogether, nearly nine hundred miles of track had been laid in Missouri by 1861. St. Louis businessmen had created the system and sold it to the state, and they had committed their city's and their own money to see it through. Small towns and rural areas had cooperated—they did not learn to hate the railroads until they were built and functioning. The state had taken on a huge debt, but it had prepared the way for access to markets and goods throughout the state. The Civil War virtually halted construction, but within a few years, Missouri railroads had opened a vast trade territory to the Southwest. Without the state treasurer's signature, the railroad bonds would not have found a market, and without the aggressive initiatives of St. Louis businessmen, the state bonds would not have been issued and the tracks would not have been laid. Governor Robert M. Stewart, a promoter of the Hannibal and St. Joseph Railroad, summed it up in 1859 when he told the General Assembly that he was "tired of the foolish cry of oppressive taxation" from those who would deny Missouri the chance "to become the Central Empire State of the Union." Without railroads, there would be no progress in Missouri. "Millions of human beings . . . will pass by us on the right and left." The great economic historian Edward C. Kirkland, in assessing the impact of railroads on midcentury economic developments, said, "Missouri undertook the greatest projects, for in the fifties the zeal for railroads put her in a ferment."[48]

---

47. Dorothy E. Powell, "History of the Hannibal and St. Joseph Railroad" (M.A. thesis, University of Missouri, 1942), 42–45.
48. Buel Leopard and Floyd C. Shoemaker, eds., *Messages and Proclamations of the Governors of the State of Missouri* (Columbia: State Historical Society of Missouri, 1922), 3:111–12; Edward C. Kirkland, *A History of American Economic Life*, 4th ed. (New York: Appleton-Century-Crofts, 1969), 387. See also Taylor, *Transportation Revolution*, 94.

Despite the twin catastrophes in 1849 of cholera, which took more than four thousand lives, and a fire that destroyed much of its commercial core, St. Louis boomed in the early 1850s. The immigrant tide from the German states waxed after the failure of republican revolutions there. Some of the newcomers were intellectuals, professionals, or military officers fleeing autocracy if not arrest, but more of them were displaced or redundant artisans attracted by the city's thriving economy. These talented people no doubt wanted to breathe free, but most of all, they wanted to do well, to prosper.

Melchior Falkenhainer, for example, was a Hessian metalsmith who came in 1852 to work as a foreman in Hudson Bridge's Empire Stove Works. His enthusiastic letters home persuaded his brother Heinrich and his friend Johanne Staehle to join him. They in turn persuaded their friend Johann Stupp of Cologne to come to St. Louis as well. The three younger journeymen had wandered for two years over the German states, Austria, and even Moscow, from one job to another. Stupp joined his friends at the foundry in 1855. A year later he opened his own machine shop on Carondelet Avenue (today's South Broadway), where he fabricated ornamental iron fences, railings, gates, religious ornaments, and the like. Thirty years later, his sons elevated the machine shop to a bridge company that became one of the leaders in the industry, and in the 1990s his great-grandsons incorporated a holding company with divisions making steel bridges, pipe for the oil and gas industry, high-rise buildings, and sugar-mill machinery. Not every immigrant was a Johann Stupp, but their presence was a vital element in the success of St. Louis.[49]

With the 4.8-square-mile city and adjacent areas crammed with people in 1855, St. Louis's limits were extended to the north, west, and south, expanding its area to 17 square miles. Some twenty thousand persons were added by the expansion, including the town of Bremen to the north. The riverfront was tripled in length to 6.8 miles and the east-west width of the city to 2.9 miles, reaching beyond Grand Avenue, the old eastern boundary of the Grande Prairie common field.[50]

In its effort to build upon its western commercial supremacy and fend off New York's and Boston's commercial outpost at the foot of Lake Michigan, St. Louis held its first Mississippi Valley Agricultural and Mechanical Fair in 1856. John O'Fallon made fifty-five acres available for the fairgrounds near the northern city limits. Exhibits of every kind of fruit and grain grown in the West were

---

49. Johanna Stupp, "Memories of My Father's Life," n.d., Stupp Papers, St. Louis (private collection); James Neal Primm, "Stupp," unpublished manuscript, St. Louis, 1999, 22–23, 195–205.
50. Primm, *Lion of the Valley,* 188.

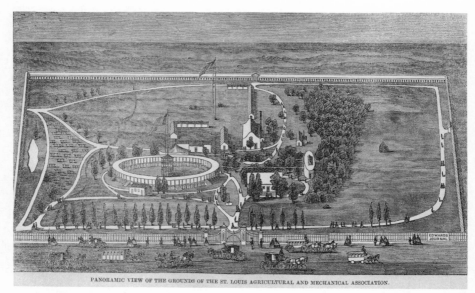

St. Louis Fairgrounds (wood engraving by Lossing-Barritt, from *Edwards Illustrated Report of the Fourth Annual Fair of the St. Louis Agricultural and Mechanical Association*, 1859). From its inception in 1856 through the end of the century, the fair provided a venue for showing off the riches of St. Louis's hinterlands and promoting its commercial interests. *Missouri Historical Society*

featured, as well as saddle- and workhorse shows and competitions for dairy and beef cattle, sheep, goats, swine, and dogs. Ten thousand persons attended the livestock shows and thoroughbred races at the arena the first year. Except for four years during the Civil War, the fair was a major event annually in St. Louis until the World's Fair in 1904—never thereafter.[51]

Yet despite the development of the railroads, despite the promise of the fair and the expansion of the city limits, the St. Louis economy proved itself incapable of sustaining uninterrupted growth. The newest troubles commenced in

---

51. In 1880, to cite one example, local newspapers reported sixty thousand fairgoers arriving at the Union Depot on the first day of the weeklong event. On the second day the rush overwhelmed the Lindell, the Southern, the Planter's House, Barnum's, the Laclede, and the smaller hotels, and crammed the trains of fourteen railroads. Premiums totaling $50,000 were awarded, and the crowds were thrilled by both thoroughbred and standardbred races, a zoological garden with African and Asian animals—a reporter noted that the lions looked bored—electric lights totaling 50,000 candlepower in the exhibition hall downtown, and a "choice collection" of European and American paintings. The exhibits attracted entrants and guests from a dozen states ranging from Ohio to Texas and the Indian and Dakota Territories. In 1878 the business leaders and publicists added another civic booster event, the Veiled Prophet celebration, held during the week before the fair. See Primm, *Lion of the Valley*, 185–86; *St. Louis Post-Dispatch*, October 6, 7, 8, 9, 1880; and Thomas M. Spencer, *The St. Louis Veiled Prophet Celebration: Power on Parade, 1877–1995* (Columbia: University of Missouri Press, 2000).

1855, when a financial crisis brought down two of the city's leading firms and threatened others. Page and Bacon, founded in 1848 by Daniel Page, the flour miller, and his son-in-law Henry Bacon, had become the largest private bank in the West, and its branch in San Francisco was the largest bank there. Its chief San Francisco competitors were Wells-Fargo and Lucas and Turner, the latter a branch of James Lucas's St. Louis bank. Both of these St. Louis firms handled large amounts of California gold, which Page was funneling into the Ohio and Mississippi. Lucas was more cautious, but he too was heavily committed to railroads. In 1854 the good times stopped rolling when New York banks, reacting to a tightening by London banks and a consciousness that railroad debt was dangerously high, began to pressure their debtors. Page and Bacon was especially vulnerable, with millions invested in railroads and in Illinois, St. Louis, and San Francisco real estate. The Ohio and Mississippi Railroad, nearing completion, was bankrupt, and Page and Bacon was running the company and paying its operating costs.[52]

At this critical juncture the Belcher Sugar Refining Company, the largest refiner in the United States, had its sugarcane plantations in Cuba seized by the Spanish government and could not redeem its notes held by Page and Bacon. The bank closed while Daniel Page negotiated a $250,000 loan from the Duncan and Sherman Bank in New York. On February 9, 1855, Page and Bacon reopened. Unfortunately, the news of its closing had reached San Francisco but news of its reopening did not. Depositors there withdrew $600,000 on the first day after hearing the bad news and continued withdrawing until the branch had to close. Duncan-Sherman, citing Page and Bacon's illiquidity, canceled the $250,000 loan. Page and Bacon assigned its assets to its creditors and closed permanently. Daniel Page was still personally rich in long-term assets, but he was no longer a dynamic force in the business community.[53]

The failure of the city's largest private bank started a run on all of the others: J. H. Lucas, the Boatmen's Savings Association, L. A. Benoist, and other banks paid out more than $700,000 in one day. On the next business day John O'Fallon, Edward Walsh, Derrick January, ex-mayor Luther Kennett (now a congressman), James Harrison, Andrew Christy, Charles P. Chouteau, and three other businessmen pledged over $8 million of their assets to guarantee the safety of the Lucas and Benoist private banks and four others, as well as the Boatmen's Savings Association. This show of solidarity reassured the public, and there were no other bank failures.[54]

---

52. Hubbard and Davids, *Banking in Mid-America*, 72–75.
53. Ibid.; Cable, *Bank of the State of Missouri*, 234–35.
54. Cable, *Bank of the State of Missouri*, 236.

Belcher Sugar's assets were seized by Page and Bacon depositors headed by Derrick January, Edward Walsh, James Harrison, and Carlos Greeley. They and others raised $100,000 in new capital, obtained a corporate charter (it had been a partnership), and restarted the flow of molasses from Cuba and Louisiana. William Belcher, the founder, who would have preferred a bailout, departed for Chicago, but his brother Charles remained as the firm's president—although without a proprietary interest. Except for the first three years of the Civil War, when it had to depend on the local supply of molasses, the firm prospered for a time. But in 1880, after successive unprofitable years, the directors fired Charles Belcher. At the time the company's assets were valued at $1,136,925. William Belcher failed in business in Chicago, and the family never forgave St. Louis.[55]

After the financial crisis of 1855, the Lucas bank was the leader on the West Coast. A former army officer, William Tecumseh Sherman, had managed the branch for Lucas since 1853. But California's volatile economy and loose bankruptcy laws alarmed Captain Sherman, and upon his recommendation James Lucas closed the branch in 1857. British interest rates were rising, reducing the flow of capital to the American railroad and markets and depleting bank reserves. Several New York, Boston, and Philadelphia banks suspended operations, causing St. Louis depositors to drain the local banks once more. The banks then pressed borrowers for payment and quit meeting short-term capital needs, forcing a number of manufacturers—including Chouteau, Harrison, and Vallé, the city's largest employer, with eight hundred workers—to close their doors. Smaller firms had to follow suit, and the panic became a depression, with thousands of workers on the streets.[56]

Trying to save the banks as before, seventeen leading citizens, including James Yeatman and John O'Fallon, pledged their assets to save the Lucas bank, but after a few days, with the signs of depression all around them, the depositors demanded their money. After paying out $1 million, J. H. Lucas and Company closed on October 5, 1857. Lucas gave his personal note to each of the remaining depositors at 10 percent interest, liquidated the bank's assets at a loss, and paid off his notes. Reportedly, this reduced his personal fortune by $500,000. As painful as this must have been, it did not impoverish him, and he barely broke stride, still the largest commercial property owner in the city.[57] By spring, the economy recovered; most of the banks resumed operations and new ones were created.

The 1857 crisis precipitated a series of changes in the regulation of the state's financial institutions. When the Bank of the State of Missouri was chartered in

---

55. Primm, *Lion of the Valley,* 193–94, 198; *St. Louis Post-Dispatch,* December 11, 1880.
56. Hubbard and Davids, *Banking in Mid-America,* 76–77.
57. Ibid.

1837 most would have preferred free banking—as many banks of issue as the economy needed. Their views hardened as the state bank denied them a more flexible currency in the 1840s and 1850s. City booster Reverend John Hogan wrote in 1854: "If we had the banking facilities possessed by cities not containing one-fourth of our population, or doing one-tenth of our business, matters would be materially changed. . . . We also need a free banking law similar to other States, and these things being had, St. Louis with her other great advantages will become the greatest of the manufacturing cities."[58]

With only the most remote farming regions opposed, the General Assembly in 1857 amended the constitution to permit ten banks of issue. The sponsors called the new system free banking, but the combined capital of the banks was limited to $20 million, and notes in denominations under five dollars were prohibited. The legislators then rechartered the Bank of the State of Missouri and created six new St. Louis banks: the Bank of St. Louis, the Merchant's Bank, the City Bank, the Exchange Bank, the Mechanics' Bank, and the Southern Bank. Each St. Louis bank could have two branches. Other banks were chartered in St. Joseph and Lexington. This situation was doomed by passage of the National Banking Act, as amended in 1865, which included measures that drove state banknotes out of circulation and made private banks obsolete.[59]

Alongside the renewed security of the monetary situation came still more people. St. Louis's population increased by 106.5 percent during the 1850s to 160,733, making it still the eighth largest city in the nation. Only Chicago and Brooklyn among the larger cities had a greater rate of increase, and only New York, Philadelphia, and Brooklyn gained more numerically. St. Louis was one of the more cosmopolitan urban counties in the nation, with 96,086 foreign-born residents, about 62 percent of them Germans, 37 percent Irish, and the remainder a scattering of English, Scots, Welsh, and Dutch. In contrast, the city's slave population declined from 2,636 to 1,542. Free blacks gained slightly, to 1,755. The flood of immigrants had made slavery redundant.[60]

Although it was still considered primarily an agricultural entrepôt and distribution center—with more than one hundred commission and forwarding houses and dozens of grocery, dry goods, and hardware wholesalers selling to retailers in Illinois, Missouri, Iowa, Kansas, Wisconsin, Kentucky, Arkansas, and Tennessee—St. Louis became an important manufacturing center during the 1850s. It was eighth in both capital invested and product value in 1860,

---

58. Primm, *Lion of the Valley,* 197–98.
59. Hubbard and Davids, *Banking in Mid-America,* 83, 161.
60. Kennedy, *Preliminary Report on the 8th Census,* 342–45; Primm, *Lion of the Valley,* 179.

more than twice Chicago's totals in these categories. Its great rival Cincinnati was still well ahead in manufacturing, with its dominance of the heavily populated Ohio, Indiana, and Kentucky markets.[61]

As in most antebellum cities, food processing led all branches of industry in St. Louis in 1860; eighteen firms were producing flour and meal valued at $4.98 million. Sugar refining produced $1.8 million, meatpacking $1.69 million and its companion industry soap and candlemaking $1.59 million, and brewing and distilling $1.76 million. St. Louisans had always been thirsty. Auguste Chouteau had operated a distillery during the Spanish period, and his descendants were still in the business, along with a half-dozen others. The brewing of beer and ale dated from 1809, but local brewers had never met the demand, even after William Lemp introduced lager in 1842. In 1860 forty breweries produced 180,000 barrels, nearly all of it for local consumption. Non-Germans quickly adapted to *lagerbier,* and beer gardens flourished all over the city.[62]

Dozens of iron foundries, two rolling mills, three steamboat and locomotive engine manufacturers, and two thriving stove manufacturers made up a metals processing industry with $2.32 million in annual product. These skilled-labor-intensive, high-value-added firms were the largest employers in the city. Other industries adding more than $500,000 to St. Louis's $27 million in manufacturing product included brickmaking, construction, men's clothing, cigar-making, and boots and shoes.[63]

Real estate values reflected the growth pattern. Real and personal property assessments, which stood at $8.6 million in 1840 and $29.7 million in 1850, totaled $102.4 million in 1860. A lot at Third and Chestnut that sold for $400 in 1826 reportedly brought $30,000 in 1858. A block bounded by Fourth and Fifth streets, Locust, and St. Charles sold for $182,000 in 1853, more than thirty times its 1833 price. A lot on Second Street, bought for $800 in 1845, was worth $142,000 in 1855. As was true of New York's first families a half-century earlier, real estate appreciation made more large fortunes in St. Louis at midcentury than did commerce and industry.[64]

The sectional crisis that crippled St. Louis business in the 1860s was already having an effect in the late 1850s. The struggle on Missouri's western border as pro- and antislavery forces battled for Kansas made eastern investors nervous about St. Louis's future. Chicago, already a desirable alternative to St. Louis because of its lakes-and-canal connection to the Northeast, gained ground with investors because of its isolation from border disputes. Some began to view St.

---

61. Primm, *Lion of the Valley,* 192–93.
62. Ibid., 192–96.
63. Ibid., 197.
64. Stevens, *St. Louis, the Fourth City,* 1:517.

Louis as southern rather than eastern-western in its cultural orientation and preferences, despite the strong Yankee element in its merchant class.[65]

When the Deep South seceded early in 1861 and the Confederacy closed the lower Mississippi River, St. Louis lost its southern and much of its eastern and foreign markets. Later in the war, its upper Mississippi trade was weakened as well, in this case by discriminatory U.S. Treasury regulations. Shipments to St. Louis were discouraged, supposedly because of its vulnerability to attack (highly overestimated), and outgoing shipments to Illinois, Iowa, and Kansas required permits from the army provost marshal at St. Louis. These rulings and delays diverted business to Chicago. On the private side, just before and during the war, eastern credit reporters judged the loyalty of St. Louis merchants to be relevant to their creditworthiness, at times basing their judgments on hearsay and innuendo. Merchants with eastern backgrounds had better luck with the provost marshals as well. Justus McKinstry, who held that office in 1861, would take a bribe, but George Leighton, Hudson Bridge's son-in-law and a McKinstry successor, was a righteous New Englander.[66]

Opinions and attitudes about the war seriously affected business relationships, especially during the early weeks. Unconditional Unionists, whose political leader was Francis P. Blair Jr., considered secession to be merely a rebellion which must be suppressed by force, while conditional Unionists believed that a state had the right to secede, that Missouri should not secede, and that force should not be used. Derrick January, president of the Merchant's Exchange, was typical of the latter group. This attitude, or perhaps some negative remarks about the Union war effort, offended some members, who seceded and organized the Union Merchant's Exchange. With the original body seriously weakened and "moderate" Unionism untenable as the war continued, George R. Taylor, president of the Pacific Railroad and originally a conditionalist himself, led those left behind into the Union Merchant's Exchange, apparently with no recriminations.[67]

Despite the army's heavy expenditures in St. Louis during the Civil War, primarily at Jefferson Barracks and at Benton Barracks—a large recruit-training facility in north St. Louis—the St. Louis economy took a heavy blow during the Civil War. There were temporary large gains for iron and lead miners and manufacturers, shipbuilders, soda-cracker bakers, millers, and meatpackers, as

---

65. Jeffrey S. Adler, *Yankee Merchants and the Making of the Urban West: The Rise and Fall of Antebellum St. Louis* (New York: Cambridge University Press, 1991), 173–74.

66. *St. Louis Missouri Republican,* October 8 and November 15, 1863, cited in Wyatt Winton Belcher, *The Economic Rivalry between St. Louis and Chicago 1850–1880* (New York: Columbia University Press, 1947), 141.

67. Primm, *Lion of the Valley,* 249.

long as the fighting lasted. With downstream markets closed by the Confederacy early in the war, and Illinois and upstream markets hampered by federal authorities, the number of commission and forwarding merchants shrank from 109 to 52, and those that survived did not prosper. Wholesale dry goods and grocery firms were limited to the army and nearby markets for their sales. Items that were not manufactured locally were difficult to obtain. After Union victories at Vicksburg and Port Hudson and the occupation of New Orleans, business improved, but when the army stopped buying in 1865, sales fell well below prewar levels. Some ingenious and adaptable businessmen, such as Hudson Bridge, prospered with alternate strategies. Bridge's agents bought Missouri cattle and hogs at bargain prices in guerrilla-threatened areas and drove them to good markets in Illinois. As the Confederates retreated down the Mississippi, Bridge's men followed, buying piles of accumulated cotton at thirty cents a pound and shipping it to a hungry New York market at seventy cents and up.[68]

After the war, with upriver markets reduced and impoverished southerners unable to buy, the Union Merchant's Exchange created a Board of Trade, which was charged with restoring St. Louis's leadership in the West. The board gathered market information, visited and established ties with market towns, and lobbied the federal government for railroad subsidies and river navigation improvements. Following Chicago's wartime example, merchants sent traveling salesmen in every direction. These "drummers" regained a little of the northern market, but as southern towns slowly recovered and St. Louis railroads penetrated the Southwest, new markets and raw material sources opened up. By 1870, twelve hundred drummers were taking orders for St. Louis goods in all of the surrounding states and in Texas as well.

Rising business opportunities were not necessarily accompanied by rising standards of business ethics. The Pacific Railroad's board of directors regained ownership of the property in 1867 by bribing Missouri legislators to surrender the state's lien on the road's assets for less than half its value. President George R. Taylor and construction engineer Daniel R. Garrison managed this affair, but the prestigious James Lucas and James Harrison and a half-dozen lesser lights went along with the $190,000 bribe. Only Hudson Bridge, an ex-president of the board, and Robert Barth opposed the bribe. The infuriated Bridge then proceeded to buy Pacific stock, including St. Louis County's large holding, until he owned the controlling interest, whereupon he exposed the whole affair to the New York bondholders and made himself president for a second time.[69]

---

68. William Ulrich to Hudson Bridge, July 17, July 25, October 10, and October 21, 1863, and H. Pearce to Hudson Bridge, November 8, November 19, November 30, December 1, December 7, December 8, December 14, December 26, and December 28, 1863, Bridge Papers.

69. See George R. Taylor Papers in general, Missouri Historical Society Archives, St. Louis. See also Primm, *Lion of the Valley*, 211–17.

In a similarly corrupt transaction, A. J. McKay acquired the St. Louis and Iron Mountain from a state-appointed commission and promptly sold it to Thomas Allen for a substantial profit. McKay was probably a front man for others, but his backers were never publicly identified. Allen was not culpable, but the dubious nature of this and the Pacific affair was consistent with Gilded Age political and business morality. Allen was determined to push the railroad to Texas, which he did, and eventually to Mexico City—which he did not.[70] Not unwittingly, Allen once again had become the city's major benefactor, playing a crucial role in its recovery and rapid growth during the next three decades. With his New York partner, Henry Marquard, Allen consolidated the Iron Mountain with two other railroads and created the St. Louis, Iron Mountain, and Southern Railroad, with connections to Memphis, Mobile, New Orleans, Charleston, and East Texas. Thus, by the early 1870s St. Louis had a huge new tributary territory to buy its products and to ship lumber and cotton to its mills, compresses, and yards. In 1880 Allen sold his Iron Mountain holdings to Jay Gould, and the road became the nucleus for Gould's five-thousand-mile Southwest System. In the same year Allen defeated Joseph Pulitzer for the Democratic nomination to Congress, whereupon the *St. Louis Post-Dispatch* inquired editorially if St. Louis was now required to think of Gould as its great benefactor.[71]

During the 1870s St. Louis led the nation in flour milling, and its meat-packing, metalworking, brewing, and saddle- and harness-making industries grew rapidly. Distilling declined as the whiskey-makers got out of town after the Whiskey Ring scandal in 1873. Two powerful industries came of age in St. Louis during the decade. First, the Liggett and Myers, Drummond, and Catlin plug and pipe tobacco manufacturers placed the city behind only Richmond, Virginia, in tobacco products.[72] Second, steamboats had carried 55,000 bales of cotton to St. Louis in 1866, primarily for Adolphus Meier's cotton mill, but as downriver and Gulf ports recovered after the war, volume declined in St. Louis. The Cotton Exchange responded aggressively by offering $5,000 for the best grade of cotton presented at the Mississippi Valley Fair in 1870. Shortly thereafter, three cotton compress companies opened in St. Louis, including the St. Louis Cotton Compress Company, the largest in the business worldwide. Its giant steam and hydraulic engines, manufactured by Gerard B. Allen's Fulton Iron Works, McCune and Gaty, and others, reduced five-hundred-pound bales of cotton to a thickness of nine inches, allowing fifty bales to be loaded onto a railroad car for shipment via the Ohio and Mississippi Railroad to eastern mills. In 1880 St. Louis was the third largest cotton market in the country and

---

70. Primm, *Lion of the Valley*, 219.
71. *St. Louis Post-Dispatch*, December 25, 1880.
72. *Report on the Manufactures of the United States at the Tenth Census, June 1, 1880* (Washington, D.C.: Government Printing Office, 1883).

the largest interior cotton market in the world. Of the 496,870 bales received at St. Louis in 1880, railroads carried 85 percent, the shippers benefiting from the Iron Mountain's special rates. In 1879 J. W. Paramore and William Senter of the St. Louis Cotton Compress Company began construction of the Cotton Belt Railroad through the Indian Territory to the Rio Grande at Laredo, opening up new cotton land to the west as it was built.[73]

The 1870 census had shown St. Louis, with 310,964 residents, becoming the nation's fourth largest city. Remarkably, it seemed, the city had doubled its population under extremely difficult circumstances. Cincinnati, Boston, Baltimore, and New Orleans were left in the dust, and Chicago staved off by a margin of 12,000. If anyone asked where 150,000 new St. Louisans had come from, the newspapers did not report it. In 1871 St. Louis annexed Carondelet and its iron smelters, and in 1875 it coaxed from the state constitutional convention a home-rule charter that split the city and county and tripled the city's area to 61.37 square miles. The expansion added several thousand people, as the Carondelet annexation had, and the immigrants kept coming, which convinced the press, public, and city fathers that all was well.[74]

To the shock and outrage of the entire city, the 1880 census showed that St. Louis had not grown at all in the previous ten years. To add to the insult, Chicago had ballooned to half a million. Protests of an undercount to the Census Bureau and its boss, Secretary of the Interior Carl Schurz (a St. Louisan himself), bore fruit, and Professor Calvin Woodward of Washington University was appointed to conduct a recount. Woodward's enumerators found 40,000 more St. Louisans, raising the total to 350,518. This was still disappointing, so Schurz invited Woodward and some of his associates to come to Washington and view the St. Louis 1870 manuscript census. Impeccably clean, without any erasures, smudges, or interlineation, it was obviously fraudulent. Not only had the population figures been adjusted to stay ahead of Chicago, but so had the economic data—so much so that industries that grew rapidly in the 1870s appeared to have declined in the 1880 census. The principal author of this fiasco had been William McKee, a central figure in the Whiskey Ring scandal and the founder of the *St. Louis Globe-Democrat*.[75]

St. Louis's actual population in 1870 will never be known, but it has been estimated at 220,000, which was enough to pass Cincinnati but not Boston, New

---

73. J. Thomas Scharf, *History of St. Louis City and County, from the Earliest Periods to the Present Day: Including Biographical Sketches of Representative Men* (Philadelphia: Louis H. Everts & Co., 1883), 2:1363.

74. Primm, *Lion of the Valley*, 272.

75. *St. Louis Republican*, October 3, October 18, and November 6, 1880; *St. Louis Post-Dispatch*, September 7 and September 11, 1880.

Orleans, or Baltimore. When the affair became public, the *St. Louis Republican* fired a parting shot at Chicago, reporting sourly that the Windy City was in a "ferment of delight" over the 1880 census results and charging that Chicago's 1880 figure had been inflated by at least 100,000.[76]

Upon reflection, St. Louisans of 1880 should have been happy. This time their city had really overtaken Boston, Cincinnati, and New Orleans, the last by more than 95,000. St. Louis was the metropolis of the Southwest, with eight trunk-line and six short-line railroads serving the city. The great Eads Bridge, completed in 1874 except for some of its approaches, was the linchpin of a vast western and southwestern railroad network. Chicago had outdistanced St. Louis in size and in the north central trade territory with the powerful assistance of the federal government during the war and Boston and New York investors pursuing their own agenda, but the "Lion" was still roaring. St. Louis had dropped to second in flour milling as the introduction of winter wheat in the Dakotas elevated Minneapolis, but was first in harness manufacturing; second in tobacco, cooperage, and sash- and door-making; third in brick- and tile-making; fifth in brewing; and sixth in iron founding, meatpacking, printing and publishing, and population.[77]

Wyatt Belcher, in his study of the economic rivalry between St. Louis and Chicago, asserted that St. Louis's leaders in the 1850s "were not willing to take risks for the sake of the development of the city and its trade area . . . [T]heir conservatism retarded the growth of the city." The reason for this timidity, according to Belcher, was that business attitudes in St. Louis reflected French traditions and methods and the "somewhat aristocratic character of the 'old families' from the South." Further, Belcher claimed that St. Louisans "subscribed to a specious economic theory according to which commerce must move along the meridians instead of East and West."[78]

All of these generalizations are wrong, despite the fact that Belcher's study is accurate in most details. The substance of the book does not support its own conclusion. On the subject of risk-taking, the roles of Thomas Allen, John O'Fallon, Daniel Page, James Lucas, and Hudson Bridge have been demonstrated here. If risking public funds qualifies, then mayor and Whig politician Luther Kennett, who helped the city spend millions on railroad stocks and bonds, was a champion risk-taker, as was the Missouri legislature. Carlos Greeley, a president of the North Missouri Railroad, was also the principal founder of and investor in the Kansas Pacific Railroad, which would connect St. Louis

---

76. *St. Louis Republican*, November 18, 1880.
77. *U.S. Census of Manufacture*, 1880.
78. Belcher, *Economic Rivalry*, 14–18, 205.

with Denver. Greeley and James B. Eads worked together on the North Missouri, and Greeley invested in and served on the board of Eads's Illinois and St. Louis Bridge Company. James Yeatman, Charles P. Chouteau, Edward Walsh, Joshua Brant, Henry Bacon, Gerard B. Allen, Samuel Gaty, James Harrison, George Collier, Henry Taylor Blow, and Adolphus Meier were all substantial buyers of railroad securities. George R. Taylor was a railroad investor as well, and he ran the double risk of being arrested for bribery—as he believed, in a good civic cause. Ironically, Pierre Chouteau Jr., though not involved in St. Louis railroads, was a major investor in the Illinois Central Railroad, Illinois' biggest project, which would connect Chicago and New Orleans—right down the meridian.

In charging St. Louis with a North-South fixation, Belcher seemed to have forgotten that Senator Thomas Hart Benton's most famous speech did not say, "there is the South, there is Patagonia." Benton was, in fact, influenced by geographer Alexander Von Humboldt's "isothermal zodiac" thesis, which stressed that the world's great civilizations had developed along a line of equal temperature, which fluctuated just above and below the fortieth parallel in the Northern Hemisphere. William Gilpin, the Bentonite editor of the *St. Louis Missouri Argus* in the early 1840s, became St. Louis's and eventually the nation's leading exponent of the isothermal zodiac explanation of cultural and economic development. In 1870 publicist Logan Reavis, who was trying to get the national capital moved to St. Louis, wrote a book entitled *St. Louis, the Future Great City of the World*, that stressed the inevitability of the vast trade of China and Europe meeting in St. Louis, which would be the U.S. center of population in the twentieth century. Of course, St. Louisans were interested in the Mississippi River and its improvements, but in the 1850s the East-West tune was far more alluring, and there was no North-South fixation, specious or otherwise. Otherwise, why build the Ohio and Mississippi Railroad and the Eads Bridge?

Belcher's third conclusion is no better than the other two. There was indeed a southern element in the St. Louis leadership, and a Creole one as well, but neither was dominant, and neither had the characteristics Belcher ascribed to them. Yankee businessmen were as numerous in the top echelons, and equally or more powerful. Thomas Allen, Henry Bacon, Sullivan Blood (president of the Boatmen's Savings Association), Oliver and Giles Filley, George Partridge, Hudson Bridge, and Daniel Page were New Englanders. New Yorkers (also Yankees to Missourians) included Henry Ames, the city's leading meatpacker; Erastus Wells, the founder of the streetcar system as well as a congressman and president of the Laclede Gas Company; and Chauncey Filley, the Republican party's leader in St. Louis from the late 1860s through the 1880s.[79]

Those southerners whom the city did count among its business leaders were Kentucky-born, with a few Virginians and Tennesseeans. Except for a somewhat more tolerant view of the Southern cause during the Civil War, they differed very little in aggressive business attitudes and methods from their Yankee counterparts, or from the James River tobacco planters of the seventeenth century or the Charleston commodity factors of the eighteenth and early nineteenth centuries. From Kentucky came Derrick January, a wholesale grocer; Wayman Crow, a wholesale goods merchant who helped found Washington University and the first St. Louis art museum; John O'Fallon; Luther Kennett; Samuel Gaty, a heavy-machinery manufacturer; John S. McCune, Gaty's partner and president of the Pilot Knob Iron Company; and James Harrison, managing partner of the Chouteau, Harrison, and Vallé Rolling Mills and a director of several railroad boards. Tennessee was represented by James Yeatman, a St. Louisan since 1842. The prestigious Yeatman was a founder of the St. Louis Mercantile Library in 1847, and a banker, iron merchant, and president of the Western Sanitary Commission during the Civil War. Virginians included George R. Taylor, a commercial builder and railroad president, and Henry Taylor Blow, president of the Collier White Lead Company and the man who developed the southwest Missouri lead district. Blow was elected to Congress as a Republican in 1864.[80]

Among St. Louis's native sons, James Lucas, Louis V. Bogy, and Louis Benoist were the most prominent in business, along with Pierre Chouteau Jr. and his son Charles P. Chouteau. Lucas was the richest person in the city, having inherited several dozen square blocks of downtown real estate. He did not rest on his oars. He had opposed coercion of the seceded states at the beginning of the Civil War, but he spent time in Washington during the war, lobbying Congress in St. Louis's and his own interests. Bogy, whose father was a Kentuckian and whose mother was a member of the Creole elite, invested in iron mining, railroads, and the Wiggins Ferry Company.[81]

Gerard B. Allen, Robert Campbell, and Edward Walsh were Irish-born; James B. Eads was a native of Indiana; Henry Shaw and builder John Withnell were English; Adolphus Meier, who founded the city's only large cotton mill, and Robert Barth were from Germany. David Nicholson, the only distiller who did not leave town after the Whiskey Ring scandal, was a Scot.[82]

79. Primm, *Lion of the Valley*, 224–26.
80. Ibid.
81. Ibid.; Edwards and Hopewell, *Edwards's Great West*, 110–13, 185–87.
82. Edwards and Hopewell, *Edwards's Great West*, 454, 508–9.

No single group, in other words, towered over St. Louis business in the immediate antebellum and postbellum years, least of all Creoles and aristocratic southerners. It is true, however, that two of the languid creditors who drove William and Charles Belcher out of the sugar refining business were southerners, while another was an Irishman who had married into the Creole aristocracy.

In his first-rate study *Yankee Merchants and the Making of the Urban West*, published forty-four years after Belcher's book, Jeffrey S. Adler agrees with his St. Louis and Missouri sources that Belcher's thesis does not scan. Adler's book stops with the Civil War, so he does not concern himself with St. Louis's postwar thrust to the Southwest. He argues correctly that slavery and the Kansas border troubles inhibited the flow of eastern capital and aggressive young entrepreneurs to St. Louis. In 1850, Adler says, St. Louis was a Yankee city, in the business sense; by 1860 it was Southern.[83] While he claims rightly that disappointed Yankees left St. Louis in droves in the 1850s, especially branch managers of eastern firms, his list is far less impressive than the list he does not make: those who stayed. Thomas Allen, Henry Bacon, Sullivan Blood, Hudson Bridge, and Carlos Greeley do not appear in his index, and Daniel Page and the Filley brothers are mentioned only incidentally. Adler seems to be interested only in the losers. New York did not shun St. Louis altogether on the eve of the Civil War, as Adler seems to imply. Hudson Bridge drew readily upon financier E. D. Morgan throughout the period and Gerard B. Allen and James B. Eads had little difficulty doing the same. Thomas Allen had access to New York funds as well, and John O'Fallon and James Lucas could have if they had wanted it.

Whatever reservations one might have about Adler's generalizations, he has done good service if he has applied the coup de grâce to the misbegotten and everlasting Belcher thesis. St. Louis's economy was, during the long lifetime of Henry Shaw, as complex and dynamic as that of any American city. The city's eventual failure to keep pace with Chicago should not mask the genius of St. Louis's entrepreneurial leaders in applying its natural, geographical advantages to the material betterment of themselves, their city, and the American West. Yet, as the endurance of Belcher's ideas makes clear, it is the St. Louis-Chicago rivalry that lives on in the historical imagination. Unproductive as that comparison may now seem, there remain at least some areas in which the resentments stirred up a century or more ago continue to justify themselves as subjects of interest. In 1875, after the St. Louis Brown Stockings won consecutive baseball games over the Chicago White Stockings, the St. Louis press crowed in delight,

---

83. Adler, *Yankee Merchants*, 173–74.

causing a *Chicago Tribune* editor to complain that St. Louisans believed that "the fate of cities had been decided by eighteen hired men."[84] So they did, and so they do—satisfied that, in one area at least, St. Louis seems sure to maintain its edge for a long time to come.

---

84. *Chicago Tribune,* May 14, 1875.

Enterprise and Exchange
The Growth of Research in Henry Shaw's St. Louis

Michael Long

When Henry Shaw first arrived in St. Louis in May 1819, the town of over four thousand inhabitants already possessed what one observer termed all the "ordinary trades and callings" and even a few "elegancies."[1] Among those elegancies was a museum—the first in St. Louis—which contained remarkable displays of Indian artifacts and natural history. The owner of the museum was remarkable, too, for he was none other than General William Clark. In 1816 Clark, who served as Governor of Missouri Territory and Superintendent of Indian Affairs, had built a special wing onto his residence to show his unique collection of Indian clothing, portraits, weapons, and tobacco pipes, supplemented with other items gathered during his administration.[2] Some things held scientific interest—among them, rare animal skins, fossil bones, and mineralogical specimens that Clark had kept from his expedition with Meriwether Lewis.[3] Clark used his private collection mainly as a diplomatic tool to impress Indian tribes during negotiations, but he also allowed other visitors to see his museum, free of charge, as long as they were respectable—which of course meant as long as they were white males, or white females escorted by proper gentlemen. Visitors who met the criteria soon learned that there was no better place for them to gain information about the frontier, and no finer collection of natural history objects west of Cincinnati.[4]

---

1. Henry R. Schoolcraft, *A View of the Lead Mines of Missouri* (New York: Charles Wiley & Co., 1819), 241.
2. John C. Ewers, "William Clark's Indian Museum in St. Louis," in Whitfield J. Bell Jr. et al., *A Cabinet of Curiosities: Five Episodes in the Evolution of American Museums* (Charlottesville: University Press of Virginia, 1967), 53.
3. John Francis McDermott, "Museums in Early St. Louis," *Bulletin of the Missouri Historical Society* 4:3 (April 1948), 130; Schoolcraft, *View of the Lead Mines of Missouri*, 241.
4. Ewers, "William Clark's Indian Museum," 54; Henry R. Schoolcraft, *Travels in the Central Portions of the Mississippi Valley* (New York: Collins and Hannay, 1825), 294.

William Clark (oil on canvas attributed to either John Wesley Jarvis or Gilbert Stuart, c. 1810; photograph by David Schultz). In addition to his roles as territorial governor and federal Indian agent, Clark founded St. Louis's first museum, which he placed adjacent to his home and office on the waterfront. *Missouri Historical Society*

Henry Shaw met the criteria, and he might have seen the collection. He was, after all, interested in the Indian trade, and a few years later came to know Clark through business. But if Shaw ever did visit the museum, he made no mention of it. His letters, journals, and receipts from the period speak mostly to the business that brought him to the city—and that in itself is revealing.[5] Shaw typified many free young men drawn to St. Louis in the decades before the Civil War. They focused on money matters and largely ignored things that did not pay. For science to grow beyond the efforts of a single individual like Clark, it would have to compete with business demands and generate the interest of more than one civic leader.

Men of science who came later to the city, like the physician George Engelmann, learned to form alliances with men of commerce and even at times to adopt business practices themselves to further their scientific ends. And men of business, like the fur magnate Charles P. Chouteau, who thought their first duty was to make money, eventually caught some of the enthusiasm for discovering and collecting new things in a country whose boundaries kept shifting westward. The meeting of these two motives was, of course, best exemplified in Henry Shaw himself, who, after making his fortune, widened his interests and dedicated a large part of his wealth to founding one of the great scientific institutions of the city. When research and business successfully intermingled, science gained a permanent cultural foothold in St. Louis.

Commerce and Curiosities

The origin of William Clark's museum could be traced, in part, to an earlier institution in Philadelphia. In 1784 the artist Charles Willson Peale had opened the nation's first museum of science and art to the public, with an avowed goal "to attract, to teach, to offer to everyone a knowledge of their world and themselves." Peale received advice and encouragement from Thomas Jefferson, and strove to display the latest information about everything, including items recently brought from the Far West. Indeed, after Clark and his partner, Meriwether Lewis, returned from their expedition in 1806, Peale was able to obtain many of their Indian artifacts and scientific specimens. He even used an exhibit to send a message to visiting Indians, displaying a waxwork figure of Lewis dressed in an Indian fur mantle and holding a peace pipe—meant to remind and reassure Native Americans that their white contact's intentions were benign.[6]

---

5. Shaw's papers are at the Missouri Botanical Garden Archives, St. Louis.
6. Charles Coleman Sellers, *Mr. Peale's Museum: Charles Willson Peale and the First Popular Museum of Natural Science and Art* (New York: W. W. Norton & Co., 1980), 1, 20, 142–43, 187.

Peale's success spawned imitators who were chiefly concerned with the commercial advantages of displaying curiosities (Clark was a notable exception). In 1829, St. Louis's first commercial museum charged twenty-five cents for visitors to see a "variety of Natural and Artificial Curiosities," including sculptures, paintings, panoramas (an early form of travelogue), and an "electric machine." The main attraction, however, seemed to be a newly invented musical instrument called the "Grand Panharmonicon," hailed as the only one in America. This instrument consisted of thirteen life-sized figures fitted with brass tongues to play bugles and trumpets, accompanied by an organ. Surely the Panharmonicon alone justified the price of admission. But after a time these novelties grew dull, and St. Louisans would seek more for their money.[7]

In 1836 they got more. It was in that year that a German immigrant named Albert Koch opened his own commercial museum in the city, and though Koch, as did his predecessors, displayed a variety of natural and artificial curiosities, he did so with a knack for both science and showmanship. For example, Koch presented not just a few stuffed birds in cases but hundreds from all over the world, in recreated habitats, next to their eggs. To supplement the usual fossil and mineral material, Koch assembled a menagerie of live animals, including at one time an anaconda, several alligators, and a grizzly bear. Koch was, in essence, the P. T. Barnum of St. Louis. Like Barnum, Koch promoted freaks of nature—an oyster shell measuring over three feet in circumference, a two-legged calf, and a two-headed, six-legged lamb.[8] He even perpetrated a hoax, a mythical creature called a "Proc," described as resembling a zebra, but with a bushier tail, stouter legs, and a rhinoceros-like horn. Koch displayed the stuffed skin of the creature and, as Barnum would later with his "Fee Jee Mermaid," invited the public to examine the specimen for themselves. He also installed waxwork figures and held evening entertainments, including everything from singers and ventriloquists to a moralistic haunted house called the "Infernal Regions," where visitors witnessed the fate of all sinners. Oddities, curiosities, entertainment, even humbug—anything to lure customers into his museum. The idea, after all, was to make money.[9]

Peale anticipated developments in curiosity museums and scientific institutions in St. Louis by several decades. For example, he charged visitors an admission price of twenty-five cents to offset expenses, sought (but did not obtain) federal government support, held musical performances at night, and promoted science as entertainment. At the start of the nineteenth century he displayed the skeleton of a mastodon—popularly called a mammoth—which started a mammoth craze.

7. McDermott, "Museums in Early St. Louis," 134–38. Another commercial museum, presumably with more science, opened in spring 1830, but closed a few months later.

8. John Francis McDermott, "Dr. Koch's Wonderful Fossils," *Bulletin of the Missouri Historical Society* 4:4 (July 1948): 233–34.

9. John Francis McDermott, foreword to *Journey through a Part of the United States of North America in the Years 1844 to 1846*, by Albert C. Koch, trans. and ed. Ernst A. Stadler (Carbondale: Southern Illinois University Press, 1972), xxi–xxii.

Yet Koch's interests ranged beyond mere income. "Mr. Koch, in his untiring efforts to please his visitors, is really a deeper practical student of natural history than he has been hitherto considered," wrote an observer in 1838.[10] What most fascinated Koch were fossils, and he used the profits from his St. Louis Museum to underwrite fossil-hunting expeditions throughout the state—efforts that made his reputation. In 1840 he found and recovered enough bones to complete a full skeleton of a mastodon, the first excavated in Missouri, which he named the Missourium. Though Koch's ignorance of comparative anatomy led him to make many mistakes in assembling the creature, he was still able to exhibit and exploit his find for profit. During one showing of the Missourium in St. Louis, for instance, he hired musicians to sit and play beneath the rib cage of the assembled skeleton while visitors toured the museum.[11] Later, however, when he took the skeleton on tour, several scientific critics in the United States and abroad charged that Koch exaggerated the size of the creature by including too many vertebrae in the skeleton. Despite such criticism Koch still was able to sell his treasure in 1843 to the British Museum for £1,300.[12] Koch's story does not end here—he eventually returned to St. Louis to contribute once more to the scientific community—but first we must see how that community came to be formed.

The First Academy:
Truth, Usefulness, and the "Golden Deity"

In 1836, the same year in which Koch opened his museum in St. Louis, a handful of professional men (mostly physicians) organized the city's first scientific society. Calling themselves the St. Louis Association of Natural Sciences, they began in a modest way to do the things that local societies did then: record and tabulate meteorological data, collect specimens for a natural history cabinet, hold meetings, read papers, and discuss the latest scientific discoveries.[13] In January 1837 they elected their first president, Henry King, who with fellow member Charles J. Carpenter belonged to the newly appointed faculty of the

---

10. *Missouri Saturday News*, May 26, 1838.
11. McDermott, "Dr. Koch's Wonderful Fossils," 248.
12. McDermott, foreword to *Journey through a Part of the United States*, xxxv; Michael J. O'Brien, *Paradigms of the Past: The Story of Missouri Archaeology* (Columbia: University of Missouri Press, 1996), 80–81, 85–86. The skeleton, reassembled more accurately, still resides at the British Museum, under its modern scientific name *Mammut americanum*.
13. Few records from this early period survive. The earliest known reference to the group comes from a table of meteorological data gathered by the association in 1836 and published in the *American Journal of Science and Arts* 32:2 (1837): 386.

Medical Department of St. Louis University.[14] Besides King and Carpenter, the most energetic members of the association were two physicians, Benjamin B. Brown and George Engelmann. Each man had a special interest. King focused on geology, Brown on zoology, and Engelmann on mineralogy, chemistry, and botany. Engelmann had by far the greatest range of expertise, and would become the most influential scientist in the city.[15]

Born in Germany in 1809, Engelmann as a young man studied botany at the Senckenberg Institut in Frankfurt before receiving further scientific training at the German universities of Heidelberg, Berlin, and Würzberg, then among the best in the world. He earned his medical degree in 1831, writing a dissertation (on plant monstrosities) that impressed Johann Wolfgang von Goethe. Engelmann numbered among his friends Louis Agassiz, who was just starting to build his scientific reputation. In 1832, after spending a summer with Agassiz and other friends in scientific study in Paris, Engelmann emigrated to the United States with the intent of investing money for an uncle and studying the natural history of the New World.[16]

On his way west Engelmann stopped at the Academy of Natural Sciences in Philadelphia, where he met Samuel G. Morton, whose large collection of human crania awed most visitors. Engelmann promised to send things of scientific interest to Morton, then continued to St. Louis, living for two years across the river with relatives and friends near Belleville, Illinois.[17] As a traveling naturalist, he made daily meteorological readings, collected plant and animal specimens, and sent duplicates back to Germany. His meteorological readings corrected Gottfried Duden's overly optimistic report of mild winters in the region, and thus gave future immigrants a more realistic picture of the climate of the Mississippi Valley. Engelmann finally decided in 1835 to make his home in St. Louis, where he opened a medical practice and quickly befriended other young men interested in science.[18]

These young professionals, catching the spirit of reform and voluntarism of the time, hoped to establish a scientific institution that would improve the intel-

14. The faculty, however, did not begin teaching until the 1840s. Nevertheless, King and Carpenter in the meantime helped build up the natural history cabinet of the Medical Department, no doubt in conjunction with their activities in the new association. See William Fanning, "Historical Sketch of St. Louis University, 1818–1908," *St. Louis University Bulletin* 4:4 (December 1908): 19–20, 27–28.

15. Engelmann was not elected president at this time, probably because he was still learning English and in general disliked making public speeches.

16. Michael Long, "George Engelmann and the Lure of Frontier Science," *Missouri Historical Review* 89 (April 1995): 251 ff.

17. Engelmann to Morton, July 28, 1837, draft copy in George Engelmann Papers, Missouri Botanical Garden Archives, St. Louis.

18. Long, "Engelmann and the Lure of Frontier Science," 255, 261, 264–65.

Dr. George Engelmann (daguerreotype by Anson, c. 1855). The German-born physician, a founder of the Western Academy of Natural Sciences and longtime associate of Henry Shaw's, was instrumental in the creation of the Missouri Botanical Garden and was perhaps the single most influential man of science in nineteenth-century St. Louis. *Missouri Historical Society*

lectual tone of the city.[19] "Men have learned," they announced in 1837, "that by associating themselves together for the cultivation of any particular branch of knowledge, they not only become of mutual assistance to each other, but by the ordeal through which they pass their opinions and discoveries . . . stamp upon them a greater evidence of truth and utility."[20] They would use their organization to guard against the dubious science sometimes presented in curiosity museums (one reason, perhaps, that Albert Koch was not among their members). They aimed at establishing peer review and higher standards within the local scientific community.

In expressing such goals, the St. Louis Association followed a national trend that began after the War of 1812. Before then, as historian George H. Daniels points out, Americans, focused largely on the tasks of nation-building, had supported only two general scientific organizations, the American Philosophical Society in Philadelphia and the American Academy of Arts and Sciences in Boston. After the war, however, Americans took a greater interest in science. Newspapers, magazines, lyceums, and popular lectures helped spread scientific information among the people. Scientific societies and journals soon flourished, and though most of them were founded in the East, some appeared in western cities like Cincinnati and St. Louis.[21]

St. Louis held a unique geographic position in relation to the advancement of scientific study in North America. When the young men of science there chose to incorporate in 1837, they looked for a new name to reflect the larger scope of their endeavor. "The attention of the scientific world," they wrote, "is everyday becoming more strongly fixed upon that immense tract of country, that lies between the western borders of civilization and the Pacific Ocean." St. Louis was located near the confluence of two great rivers—the main system of transport on the frontier. The city already played a crucial role in commercial transshipping. Steamboats had to load and unload cargo at St. Louis to transfer goods between the heavier craft plying the lower Mississippi and the lighter boats on the upper rivers. A scientific society could exploit this hub of exchange, acting as an outfitting station for expeditions heading west as well as a clearinghouse for specimens and information coming east. Thus the members were poised to investigate the West better than any other institution in the world, and the

---

19. A survey of contemporaneous city directories shows that many of these professionals belonged to a variety of civic organizations in St. Louis—for example, the German Benevolent Society, the Franklin Society, and the St. Louis Library Association.

20. *Act of Incorporation, Constitution and By-Laws of the Western Academy of Natural Sciences at St. Louis* (St. Louis: William Weber, 1837), 13. Engelmann's copy is at the Missouri Botanical Garden Library.

21. George H. Daniels, *American Science in the Age of Jackson* (1968; reprint, Tuscaloosa: University of Alabama Press, 1994), 9, 13–14.

name they finally chose—the Western Academy of Natural Sciences—let the world know it.[22]

Even before the state legislature granted the organization its charter, the Western Academy launched an ambitious program to prove its worth to Missouri and the nation. Since many of the members were physicians, they knew the importance of accurately observing meteorological conditions (temperature, barometric pressure, rainfall, thunderstorms, etc.) which, according to medical theories of the day, directly affected health in the community. Academy members also knew that many Americans desired not only health but wealth from their land. Starting in January 1837 the academy offered free analysis of any mineral sample sent to them from the region, provided the sender paid for shipping. In exchange for its analysis, the academy kept the sample and enlarged its mineralogical collection.[23] Members also gathered their own specimens when they could. In February 1837 the mineralogy and chemistry department, headed by George Engelmann, discovered a three-foot vein of anthracite coal in Washington County, Missouri, in the Iron Mountain mining region.[24] Engelmann expanded his search in March when, taking advantage of the slow season in his medical practice, he accepted a commission to travel to Arkansas to test minerals there for precious metals. Along the way he toured more of the mining regions and collected mineral and plant specimens for the academy.[25]

Though they readily tapped into the common desire among westerners for easy wealth, the members of the academy also hoped to elevate men's spirits above monetary concerns. The Unitarian minister William Greenleaf Eliot, an early member of the Western Academy, emphasized the moral advantages that such an organization as the academy could bring to rough-edged St. Louis. In Eliot's view, the West—though possessed of its share of scoundrels and idlers—attracted predominantly honest, hard-working young men who nevertheless stood in danger of losing sight of the spiritual and intellectual virtues of civilization. These men could benefit most from civic organizations like the Franklin Society (a local literary group) and the Western Academy, which would be the means "by which public opinion may be elevated, public feeling and taste purified, and the community saved from forgetting that there is something real in the world, besides money." Eliot saw too many of his fellow citizens worshipping the "golden deity."[26]

22. *Act of Incorporation,* 13, 14.
23. *St. Louis Missouri Republican,* January 13, 1837.
24. *Anzeiger des Westens,* February 4, 1837.
25. Jerome Jansma and Harriet H. Jansma, "Engelmann Revisits Arkansas, The New State," *Arkansas Historical Quarterly* 51:4 (Winter 1992): 334–35, 339–40.
26. William G. Eliot, *Address Delivered before the Franklin Society of St. Louis on the Occasion of its First Anniversary, January 7th, 1836* (St. Louis: Charless & Paschall, 1836), 23, 20.

Ironically, this was the very deity that the Western Academy needed most to appease. The academy's members were not paid to study science. They still had to earn a living like other citizens. And they soon found that the money they earned was not enough to support the academy in all its goals—indeed, not even in some of its necessities. Renting a meeting hall was at first beyond their means. Nor could they afford to house their collections adequately. The departments of zoology, mineralogy, geology, and botany bulged with specimens—the herbarium alone contained thousands of plants from the region collected by Engelmann—but for lack of space, the members had to store items separately in their homes.[27] (This arrangement made it difficult for anyone to study the natural history cabinet as a whole—one of the ostensible advantages of forming an academy in the first place.) Even when they finally did rent a hall, they could not display large mammals. Nor could they afford to buy natural history specimens from other collectors as they became available.[28]

To help relieve this situation, the members petitioned the Missouri legislature, and later the U.S. government, for financial aid. Both governments, however, denied their requests.[29] The Western Academy thus would have to count on membership dues and patronage for support, and hope that the zeal of its members would carry it through tough times. Though the organization struggled at first, there were signs even in the first year that its fortunes might rise. By September 1837 the academy had seventeen active members, including sixty-five-year-old Judge Marie P. Leduc, who from the start let the academy meet in his office and served as the group's treasurer. Colonel John O'Fallon (William Clark's nephew) helped the academy in 1837, as did the colorful Scotsman William Drummond Stewart, who was passing through St. Louis on his way west for adventure. By year's end the academy could afford to rent a meeting hall and display some of its collections.[30] Those collections grew in prestige when William Clark, through his son Meriwether, generously donated to the academy the scientific portion of his museum.[31] By this act the Clark family acknowledged the promise of the organization and their belief that the academy could best ensure the scientific legacy of Clark's museum.

---

27. *Missouri Saturday News*, February 10, 1838.
28. *Act of Incorporation*, 3, 14.
29. Walter B. Hendrickson, "The Western Academy of Natural Sciences of St. Louis," *Bulletin of the Missouri Historical Society* 16:2 (January 1960): 117.
30. *Missouri Saturday News*, February 10, 1838.
31. John Francis McDermott, "William Clark's Museum Once More," *Bulletin of the Missouri Historical Society* 16:2 (January 1960): 131. As McDermott shows, Clark at the same time gave many of his Indian artifacts to Albert Koch and his museum.

The Reach of Reputation

The status of the Western Academy also grew among prominent men of science here and abroad. In 1837 the French mathematician, astronomer, and cartographer Joseph N. Nicollet visited St. Louis, on his way west with his young assistant John C. Fremont to explore and map the land between the Mississippi and Missouri rivers.[32] Nicollet formed a working relationship with the academy that aptly demonstrated the role the organization could play in western exploration. Academy member George Engelmann, for instance, helped outfit Nicollet with scientific equipment—a chisel punch, hammers, and chemical tests for geological surveys, as well as vaccine matter to prevent the spread of smallpox among Indians.[33] When Nicollet needed a plant collector for his expeditions, academy member Karl Geyer signed on for both the 1837 and '39 trips. Most of all, Nicollet needed conscientious observers to record daily barometric pressure readings in St. Louis, which he then compared to his own readings in the field to help determine altitudes for his map; Engelmann and the academy steadfastly provided this service.

In return, Nicollet supplied Engelmann with a new barometer to carry out more accurate readings, gave a copy of one of his scientific reports to the academy, and personally carried Engelmann's first botanical manuscript back east for publication in the *American Journal of Science and Arts*.[34] Nicollet, a corresponding member of the academy since 1837, hoped to return to St. Louis after he finished his map, to aid the academy even further.

The Western Academy also formed international connections. The members sent the results of their 1837 winter observations to the English astronomer John Herschel (at the time in South Africa), having followed Herschel's recommendation of making meteorological readings for twenty-four hours on the winter solstice.[35] Through Engelmann (who kept in touch with German friends, particularly botanists Alexander Braun and Georg Fresenius), the academy sent to the Senckenberg Institut in Frankfurt a half-dozen boxes filled with enough natural history specimens to start a small museum: deer, minx, raccoon, muskrat, fox, eagle, owl, prairie chicken, turtle, mussel, salamander, frog, lizard,

---

32. The best recent study of Nicollet and his contributions to science is Martha C. Bray, *Joseph Nicollet and His Map*, 2d ed. (Philadelphia: American Philosophical Society, 1994).

33. Donald Jackson and Mary Lee Spence, eds., *The Expeditions of John Charles Fremont* (Urbana: University of Illinois Press, 1970), 1:42, 77.

34. Nicollet to Engelmann, June 7, 1841, Engelmann Papers; Henry King and Engelmann to Nicollet, November 16, 1837, Nicollet Papers, Minnesota Historical Society Archives; Engelmann to Asa Gray, November 19, 1841, Historic Letters Collection, Gray Herbarium Archives, Harvard University, Cambridge (hereafter cited as GHAH).

35. *Missouri Saturday News*, February 10, 1838.

forty-three fish, six hundred insects, and over a thousand plants. Among the more impressive specimens, surely, were the skull of a two-year-old black bear and the entire skeleton of a young bison.[36]

In return, the academy hoped to receive European specimens from Senckenberg. The academy asked not for a complete collection, but simply enough to give Americans an idea of those species for comparison. It was normal for men of science to conduct business this way, especially with colleagues abroad, because specimen and publication exchanges simplified transactions. But the Western Academy hoped for something more than an equal exchange. The members considered their organization an offspring of the Senckenberg Natural History Society, and thus appealed to the parent organization for indulgence. Send good specimens, they told Senckenberg, regardless of cost, because "in furthering science, the rich should give to the poor when they can." Though they had no publications to trade, they still wanted their European brethren to send them "books and books and books."[37]

Despite their lack of books, the members of the Western Academy were asked to help revise one. In 1840 John E. Holbrook, the father of American herpetology, requested a loan of all the academy's reptile specimens, especially rattlesnakes, to aid him in preparing a new edition of his seminal study of North American reptiles and amphibians.[38] Engelmann was initially willing to bend the academy's rule about not loaning out its permanent specimens. Perhaps, though, he either thought better of his offer or was overruled by the other members, because Holbrook does not mention the Western Academy in the revised edition of his work.[39]

By 1840 the Western Academy seemed sure of success. Even in the midst of a national financial panic, the members acquired five acres near the city in which to plant a botanical garden. They also continued to add to their natural history cabinet. Though George Engelmann left them early in the year for a marriage trip to Europe, he continued to act on his journey as an ambassador of science. He visited colleagues in Philadelphia, Frankfurt, Berlin, Göttingen, Prague, Warsaw, and Vienna; collected human crania in Germany for Samuel G. Morton; carried books for Morton overseas; and explored the possibility of translating the work of a German scientist into English.[40] No doubt everywhere

---

36. Adolph Reuss and Engelmann to the Senckenberg Natural History Society, May 20, 1838, Senckenbergischen Naturforschenden Gesellschaft Archiv, Frankfurt-am-Main, Germany. (In June 2000 Dr. Konrad Klemmer of the Senckenberg Museum graciously read the old German script in my presence while I wrote down his translation.)
37. Ibid.
38. Holbrook to C. J. Carpenter, June 13, 1840, Engelmann Papers.
39. John E. Holbrook, *North American Herpetology*, 5 vols. (Philadelphia: J. Dobson, 1842).
40. Engelmann to Morton, August 22, 1840, copy in Engelmann Papers.

he traveled he informed his friends about the Western Academy and its hope for the future. He hardly could have seen the odd mix of failure and success that lay just ahead.

Hard Times

During the 1840s science in the United States gained national momentum. The decade began with the birth of the National Institute, a large natural history cabinet whose premature demise made way for the Smithsonian Institution in 1846. That same year the Swiss scientist Louis Agassiz arrived on the East Coast and began captivating audiences with his lectures, while in the Southwest the United States launched a war against Mexico to capture more territory in the name of Manifest Destiny. New territory required exploration and surveys to fix borders; such exploration led, in turn, to an explosion of new species discoveries and other scientific information coming from the West. At war's end in 1848 the scientific community took the next step in professionalization and founded the American Association for the Advancement of Science.

Yet by the middle of the 1840s the one scientific academy in St. Louis, which should have profited from all this activity, was failing. The members had to disband their collections, store them separately in their homes, and meet only informally as occasion required. What went wrong? Historian Walter B. Hendrickson argues that the academy ceased operations largely because it lacked adequate funding, and because within a few years the group lost several key figures.[41] What he does not mention is that the busy careers of the remaining members, coupled with the growth of western exploration, worked to paralyze the academy. The many scientific specimens coming from the west overwhelmed someone like George Engelmann, who could study them only in his spare time. There simply was too much science spread among too few practitioners.

The first important academy member to leave St. Louis was Henry King, who moved to the East Coast in 1839. King's relocation might have yielded good results for the academy, if not for political interference. Through his Washington friends King became a curator for the National Institute, and he wrote to Engelmann about the collections arriving daily from the Wilkes Expedition. Engelmann wanted to obtain specimens from this famous survey of the Pacific Northwest and beyond, but King was unable by law to exchange anything with

---

41. Hendrickson, "Western Academy," 129; Walter B. Hendrickson, "Science and Culture in the American Middle West," *Isis* 64 (September 1973): 330.

the academy, and was himself removed from curatorship after a short while because of political wrangling. Thus the academy lost both a founder and a potential source for new holdings.[42]

Western exploration and death contributed to the demise of the academy as well. Adolph Wislizenus, Engelmann's medical partner, joined a Santa Fe caravan in 1846 at the start of the war with Mexico, and Benjamin B. Brown headed to California during the gold rush years. In 1842 Judge Marie P. Leduc, who had served as treasurer for the group since its beginning, died at age seventy. The next year Joseph Nicollet, at work in Washington, fell ill and died before completing his map.[43] Had Nicollet been able to return to St. Louis as he had hoped, he might have injected new energy into the failing organization.

Economic depression also hastened the end of the academy. When the effects of the Panic of 1837 finally reached St. Louis, even normally solvent men like Pierre Chouteau Jr. and Meriwether Lewis Clark had to turn to wealthy men like Henry Shaw for loans.[44] Academy members were particularly vulnerable. Many were physicians, still establishing their careers or starting families, and though medical men were just as busy then as now, most did not grow rich from medicine alone. Physicians often invested in property, which lost value in a depression. Some academy members may not have been able to pay their annual dues. The academy had to sublet its hall to the Medical Society of Missouri in 1842. The income—a mere three dollars a night, which included firewood and candles—could not stave off the inevitable. By 1843 the Western Academy stopped subletting and vacated its own hall.[45]

Botany as Business Venture

Though the Western Academy slowly shut down its operations, scientific research continued in St. Louis. The scientists who remained in the city narrowed their focus to favorite specialties. Engelmann, for instance, devoted most of his spare time to meteorology and botany. Since 1840 he had corresponded with Asa Gray, the leading botanist in America, exchanging letters and plant specimens that helped both men. Engelmann early on revealed a businesslike knack

---

42. King to Engelmann, May 9, 1842, and January 14, 1843, Engelmann Papers.
43. Bray, *Joseph Nicollet and His Map*, 292.
44. Henry Shaw account book, July 6, 1840 (Journal 1839–1853, 10:155), Henry Shaw Papers, Missouri Botanical Garden Archives.
45. St. Louis Medical Society Constitution and Minutes, April 1, 1842; May 5, 1843; September 1, 1843, typescript copy in Archives and Rare Books, Bernard Becker Medical Library, Washington University, St. Louis.

for maneuvering within the mail system when, to avoid unreasonable postal charges, he urged Gray to hide pressed plants inside newspapers and pamphlets before sending them west—a practice that meant paying higher rates only if the plants were discovered.[46]

By January 1843 Engelmann showed even greater business acumen when he thought of a way to send more western plants to Gray. Two of Engelmann's German-American friends, Karl Geyer and Jacob Lindheimer, wanted to earn their livings as collectors. Geyer already had traveled with Nicollet and now hoped to join one of the spring caravans to Oregon. Lindheimer wished to concentrate his efforts in Texas. Because both men collected and preserved good specimens, Engelmann saw possibilities. "I have conceived the idea to offer their collections for sale," he wrote Asa Gray at Harvard. "We would guarantee the genuineness, the good preservation and careful selection of the specimens." In other words, Engelmann and Gray would ensure that the plants collected were new species, not merely claimed as such to make more money for the collectors. "What do you say to that plan?"[47]

Gray said yes, and the two men formed a plant partnership. Gray needed western specimens to complete *Flora of North America*, which he was writing with John Torrey. Engelmann needed Gray's expertise and resources (eastern libraries and herbaria were more complete than those in St. Louis) to confirm new species. All monies would go to pay the collectors and to cover operating expenses. Neither Gray nor Engelmann would earn anything from the venture except an increase in their scientific reputations and the knowledge that they were providing a valuable service. For as immigrants headed west, they brought seeds, some carried intentionally, others hitching rides in grain sacks and boxes or on the mud of wagon wheels and animal hooves—seeds of opportunistic plants from the East and Europe that would invade the native flora and make it "more difficult to decide which are natives and which introduced."[48] Thus the urgency to discover and record what was indigenous in the West before invading flora supplanted native species forever.

By early spring of 1843 the great migration to the Oregon Territory had begun. St. Louis teemed with visitors—immigrants buying last-minute equipment and supplies, Santa Fe traders preparing to make their yearly trek to the

---

46. Engelmann to Gray, December 14, 1841, GHAH.
47. Engelmann to Gray, January 18, 1843, GHAH.
48. George Engelmann, "Catalogue of a Collection of Plants Made in Illinois and Missouri, by Charles A. Geyer," *American Journal of Science and Arts* 46:1 (1843): 94 n. Historian Alfred W. Crosby shows that the plant invasion was already well advanced in California by the eighteenth century; see his chapter "Weeds" in *Ecological Imperialism: The Biological Expansion of Europe, 900–1900* (Cambridge: Cambridge University Press, 1986).

Southwest, soldiers and adventurers bound for new lands. The man sent by the army to map a new route to the Pacific Coast was John C. Fremont, who took with him about forty men, a small cannon, and a wagonload of scientific instruments. Fremont met with George Engelmann to coordinate their barometers for determining altitudes along the trail.[49] The naturalist John J. Audubon also stopped in St. Louis on his way up the Missouri River to study birds and quadrupeds. He received Engelmann and a small deputation from the Western Academy bearing a diploma of honorary membership, which pleased him.[50] The Scotsman Sir William Drummond Stewart arrived in town for a fifth western adventure. Stewart sought to command a party of nearly fifty men, and though he failed to persuade Audubon to join the caravan, he welcomed Karl Geyer, along with another man Engelmann hired for the plant business, Friederick Luders. Geyer and Luders planned to branch off eventually from Stewart's party and travel on their own.

As in all business ventures, the Engelmann-Gray plant enterprise ran risks. Lindheimer in Texas, for example, suffered bad luck when illness, rain, and Indians delayed him from broaching the interior. Luders traveled all the way to the Columbia River before the swift current of that waterway overturned his canoes and swept away his entire collection. The worst setback, however, came from Geyer. Engelmann had taken a surety on a loan to outfit Geyer, risking $150 of his own money to do so. Geyer signed a contract, headed west, collected many excellent specimens, but then, instead of returning to St. Louis, boarded a ship on the Northwest Coast and landed eventually in Europe, making excuses all the way. Engelmann lost his money and any chance of recovering it from the sale of Geyer's specimens.[51]

Despite these setbacks, the plant partners were able to sell many specimens that Lindheimer eventually collected in Texas. Subscribers to the collection included both individuals and institutions in the United States and abroad. Some subscribers were physicians like Charles W. Short of Kentucky, who wished to enrich a private collection. Others were men attached to scientific institutions, like Joseph Decaisne in Paris and G. Bischoff in Heidelberg. In time the list of institutions receiving plants from Engelmann and Gray included the

---

49. As did his mentor Nicollet, Fremont relied upon Engelmann's barometric readings to calculate altitudes in the West. See Jackson and Spence, eds., *Expeditions of John Charles Fremont*, 1:317–37, 372, and map no. 3 in the Map Portfolio volume.

50. Maria R. Audubon and Elliott Coues, *Audubon and His Journals*, vol. 2 (London: John C. Nimmo, 1898), facsimile of diploma, appendix (n.p.); John Francis McDermott, ed., *Audubon in the West* (Norman: University of Oklahoma Press, 1965), 14, 60–61.

51. Engelmann to Gray, November 22, 1843, GHAH. See also Geyer's letters to Engelmann from this period, Engelmann Papers.

Academy of Natural Sciences in Philadelphia, the Smithsonian Institution, the Royal Botanic Gardens at Kew, and the British Museum.[52]

An even greater opportunity for collecting came in the spring of 1846, when Engelmann's medical partner, Adolph Wislizenus, yielded to wanderlust and joined a caravan headed for Santa Fe. Though the United States was on the brink of war with Mexico, Wislizenus agreed to collect in Mexican territory for Engelmann, as did Josiah Gregg, the celebrated author of *Commerce of the Prairies*. After war was declared, Engelmann and Gray seized the opportunity to launch a third man, August Fendler, into the field by getting him permission to travel with a group of U.S. soldiers to the Santa Fe region. All these efforts paid scientific dividends. Though Mexican soldiers captured Wislizenus, they placed him under house arrest in one of the most botanically rich areas of northern Mexico, where he collected for many months. Gregg and Fendler escaped capture and also made valuable collections.

These collections brought Engelmann many new specimens to study and describe. Early on he had developed an interest in cacti, and now became an authority on them. After the war, more collectors went west with military topographical units, which were mapping the new boundary between the United States and Mexico. Novel species flooded eastward to Engelmann, Gray, and Torrey, and overwhelmed them. Engelmann had the increased pressure of a successful medical practice. Not only did his patient load instantly double when his partner Wislizenus left, but cholera struck St. Louis hard in 1849, killing some four thousand and subsequently returning each summer, though with less devastation, for several years afterward. Engelmann had botanical reports to write, patients to see, and plant specimens to study. At times he imagined himself heading west to collect and observe new plants in their habitats. But in the end he left the remote collecting to the government-sponsored parties or the rare individual setting out on his own.

One person he enlisted as a collector was Ferdinand V. Hayden, a young geologist and paleontologist who arrived with a colleague in St. Louis in 1853. The two young men were on their way to explore the Badlands in Nebraska Territory, but when another group of scientists, backed by the governor of Missouri, objected to the competition, George Engelmann and Louis Agassiz (who was visiting Engelmann at the time) were called in to mediate. They spoke on behalf of the young men, reminding the group that the West was large enough to accommodate as many scientists as could travel there. Eventually the young men

---

52. George Engelmann Herbarium: Desiderata and exchange lists, 1838–43, 1844–45, folders 13–14, box 48; Herbaria and Collectors: August Fendler, subscribers to Fendler's plants, folder 6, box 50, Engelmann Papers.

were allowed to share the one available boat with the state-sponsored party, but the incident left Engelmann disheartened. "There is want of the true spirit of science, the pure love for science in all this," he confided to James Hall in New York, who was sponsoring the young traveling scientists.[53] Engelmann could not foresee that three years later Hayden would help ignite the "true spirit of science" again in St. Louis.

New Patrons, New Institutions

St. Louis more than doubled its population in the 1850s, growing from nearly 78,000 at the start of the decade to over 160,000 by the end. The city could now support more institutions like those William Greenleaf Eliot had hoped would enlarge men's minds (women were excluded from the benefits of such institutional enlargement until the latter part of the nineteenth century).[54] Two medical schools flourished in healthy (or perhaps unhealthy) competition, while a polytechnic school and a new institution of higher learning, Washington University (which Eliot helped to start), arose during the decade. Furthermore, the city now contained a handful of commercial men genuinely interested in science. For James B. Eads, who earned a living salvaging wrecked steamboats, the interest grew out of practical engineering experience. For Charles P. Chouteau, who grew up in the fur trade, the interest seems more surprising. Yet Chouteau was deeply in earnest about science, pledging his own money, labor, and resources to advancing the cause.

As the son of Pierre Chouteau Jr. and partner in the family business—locals still called it the American Fur Company—Charles Chouteau was well placed to help men of science, particularly in their travels west. Charles's father had arranged passage for John J. Audubon on a company boat in 1843, and Charles kept up the tradition in 1854 when he allowed Ferdinand Hayden to travel upriver again to Nebraska Territory. In return for offering Hayden free passage

---

53. Quoted in John M. Clarke, *James Hall of Albany: Geologist and Palaeontologist, 1811–1898* (Albany, N.Y.: n.p., 1923), 248–49.

54. Two exceptions are worth mentioning. In 1850 the young sculptor Harriet Hosmer managed to study at the McDowell Medical College, but only to improve her artistic understanding of anatomy. Women throughout the country were generally barred from formal medical studies, a practice that also denied them access to the sciences—medicine being the main way to gain scientific training. After the Civil War, women played more active roles in scientific inquiry. The St. Louis entomologist Mary Murtfeldt, self-taught and brilliant in observing insect life cycles, by the turn of the century had published many of her findings, been elected an honorary member of the Academy of Science of St. Louis, and served eight years as the Acting Missouri State Entomologist.

Charles P. Chouteau (steel engraving by American Biographical Publishing Co., n.d.). Chouteau used his considerable family interests in the fur trade to support his own interest in science and exploration. Like his father, he backed scientists traveling up the Missouri River; additionally, he purchased collections for public display and study in St. Louis. *Missouri Historical Society*

and the hospitality of company forts along the way, Charles became part owner of Hayden's collection, taking possession of one-fourth of the fossil specimens. This business arrangement eventually triggered an event that reawakened the scientific community to the need for an active scientific association in St. Louis.

Hayden collected in Nebraska Territory for two years and returned to St. Louis during the bitter cold of January 1856, bringing with him an immense collection of natural history specimens—six tons of fossils, marine shells, reptiles, fishes, bird skins, and plants. The fossils alone weighed over a thousand pounds and included "the first soft-shelled turtle known in America," dinosaur teeth, and "a magnificent collection of tertiary quadrupeds," including rodents, peccaries, hippopotami, and a small, three-toed equine called an Hipparion—only the second one found in the United States. Chouteau immediately took charge of the entire collection, setting it up for public display in his house at Sixth and Olive, where a local reporter urged all scientific men to visit.[55]

Men of science responded with enthusiasm. Hayden's specimens impressed them so much that they wondered how to keep the whole collection—not just Chouteau's portion—in St. Louis. Should they revive the old Western Academy? Or did they need a new organization? Finally, in late January fifteen men, including Charles Chouteau and George Engelmann, met to discuss starting a new scientific society. Though other responsibilities weighed heavily on Engelmann (he had to finish botanical reports, establish a new hospital, and prepare for a second European trip), he nevertheless agreed to chair the first meeting on February 8.[56] In his opening remarks Engelmann raised two concerns: the new society needed to collect enough money to survive, and the members needed to work towards obtaining the valuable fossils that lately stirred their imaginations. On the second matter they received generous help when Charles Chouteau donated his portion of the fossils to the new organization, now called the Academy of Science of St. Louis.[57]

Chouteau went even further. Before leaving on a trip that spring to the upper Missouri River, he asked the academy to provide him with "suitable antiseptics" to distribute among the "various trading Posts on the Upper Missouri for the preparation of specimens in Natural History."[58] When he returned a few months later, he brought the academy new specimens of antelope, fox, and other animal skins, along with fossils and Indian artifacts.[59] The next year Chouteau planned to travel up the Missouri River and employ a taxidermist—at no charge to the academy—to properly collect and prepare specimens on site. He would also take a small seine to drag streams along the way and gather plenty of new material for the museum.[60] In other words, Chouteau committed part

55. Mike Foster, *Strange Genius: The Life of Ferdinand Vandeveer Hayden* (Niwot, Colo.: Roberts Rinehart, 1994), 66–67; Engelmann to Gray, January 29, 1856, GHAH; *St. Louis Missouri Republican,* February 4, 1856.
56. Engelmann to Gray, February 8, 1856, GHAH.
57. Minutes of council meetings of the Academy of Science of St. Louis, February 8, 1856, and April 21, 1856, Missouri Botanical Garden Library, microfiche.
58. Ibid., May 2, 1856.
59. Journal of Proceedings, August 4, 1856, in *Transactions of the Academy of Science of St. Louis* 1: 21–22, Missouri Botanical Garden Library.
60. Minutes of council meetings of the Academy of Science, March 23, 1857.

of his commercial network and resources—even his own labor—to strengthen the academy's reach into the West. Over the years that commitment would add greatly to the animal, mineral, and ethnographic specimens in the academy's museum.

Not all men of business, however, possessed Chouteau's generous spirit. One even tried to take advantage of the academy. From the beginning the new organization needed a place to meet and display its collections. Edward Wyman, owner of the Odeon Hall, where Jenny Lind had sung to adoring audiences in 1851, made the academy what seemed at first an attractive offer. For years Wyman had run a successful private school at the Odeon, keeping a museum of science and art on the upper floors of the building for the benefit of his pupils. In 1853 Wyman retired as an educator and devoted his time to commercializing his museum. That same year he bought from Albert Koch the fossil skeleton of a giant "sea serpent" called a Zeuglodon, which no doubt attracted paying customers.[61]

By 1856 Wyman wished to retire. He offered for "cash, or nearly so," the entire collection (which, besides the Zeuglodon, included the usual shells, minerals, and birds) to the academy for $20,400. When members balked at the price, Wyman made a second offer. The members of the academy could use his collection—and rent rooms at his hall to display their own specimens—for $1,200 a year. Wyman would pay all operating expenses, and though the academy would retain control of its collection, Wyman would have the exclusive right to open the hall to the public and charge an admission fee for special viewings.[62]

Though combining the collections would have instantly increased the prestige of their museum, academy members wisely rejected the offer as "wholly inadmissible and impracticable."[63] Not only would Wyman retain the right to end the arrangement at any time, leaving the academy homeless, but he also would earn money from showing the academy collection in his hall—while charging the academy rent! The academy members decided instead to keep the temporary arrangement they had worked out with physician Charles Pope, a founding member. Pope headed the St. Louis Medical College and he was able to offer the new academy a free room at the dispensary, part of a building his father-in-law, John O'Fallon, had financed. The academy stayed there for free, courtesy of Pope and O'Fallon, for thirteen years.

Though the members still wanted a Zeuglodon skeleton for their museum, they sought a cheaper way to obtain one. They commissioned Albert Koch,

61. McDermott, foreword to *Journey through a Part of the United States*, xxxi.
62. Minutes of council meetings of the Academy of Science, March 10, 1856; April 21, 1856.
63. Journal of Proceedings, June 2, 1856, in *Transactions* 1.

once again living in St. Louis, to travel to the state of Mississippi, where another such creature was rumored to have been found, and to scout the possibility of excavating it. Koch reported back in June that all the fossils could be purchased for about $200, but in the end the academy decided not to buy the skeleton. They did accept "a large and valuable collection of Cretaceous and Tertiary fossils," however, which Koch gathered on the journey and offered to donate to the organization.[64] Like its predecessor, the academy would grow more through donations and exchanges than outright purchases.

Another Plant Partner

The spring of 1856 marked the beginning of another scientific institution in St. Louis. Henry Shaw, by now a St. Louisan for nearly forty years, had done well in his hardware business and in subsequent real estate transactions. Now semiretired, he wished to devote part of his wealth to transforming his country estate into a public garden. He would finance the project alone. Still, he needed help in refining and carrying out the plan, so in March Shaw wrote to Sir William Hooker, head of the Royal Botanic Gardens at Kew, for advice. Hooker wrote back and encouraged Shaw to make his garden scientific. By good fortune, the best person to help him was close at hand—a longtime correspondent of Hooker's, George Engelmann. Shaw at once got in touch with Engelmann, and thus began an association that eventually led to the founding of one of the world's great botanical gardens.[65]

Whether Shaw knew it or not, by turning to Hooker he had entered a realm of conspiring scientists. Hooker, Engelmann, and Asa Gray spent the next few years corresponding with each other to discuss how to bend Shaw to their collective scientific will. Engelmann, their man-on-the-scene, learned to his dismay how difficult this was. After initial contact with Shaw that spring, Engelmann confessed to Gray that he had "not yet seen much of Shaw, and am unfortunately not the proper person in address and diplomacy . . . to work upon him." Shaw, though sufficiently energetic, was also not the right person in Engelmann's view to carry out the project, because he lacked education in science as well as the taste for it, and often refused to alter his plans to accommodate the needs of botany.[66]

---

64. Ibid., April 21, 1856.
65. For more background on Shaw and his garden, see William Barnaby Faherty, *Henry Shaw, His Life and Legacies* (1987; reprint, St. Louis: Missouri Botanical Garden, 1993).
66. Engelmann to Gray, May 13, 1856; April 15, 1859, GHAH.

Henry Shaw (from stereoscopic photograph by Emil Boehl, c. 1880s). Shaw, an amateur botanist and professional money-maker, exemplified the entrepreneurial spirit that underlay the development of scientific research in St. Louis. *Missouri Historical Society*

Over time Engelmann became friends with Shaw, but their relations still occasionally suffered from the clash of ideas. "I am afraid we can not pull well together—I can not influence him," Engelmann complained to Gray. He did not know how to flatter Shaw. "A man who has no real scientific zeal nor knowledge[,] who must be got to do things by diplomacy, I can not do much with. The proper way would be to get him interested in what interests us and seems important to us, but that I unfortunately do not understand."[67] The trouble

67. Engelmann to Gray, April 10, 1860, GHAH.

lay partly in different temperaments. Engelmann was a man of perpetual motion. His medical practice and scientific investigations consumed his waking hours, and he was impatient with men who operated more leisurely. He once jokingly asked a friend if some of the fellows who always seemed to have too much time on their hands could "make me a present of half or three fourths of theirs?"[68] Shaw, by contrast, lived at a slower pace. While he acknowledged the need to build a museum for plant specimens and a library (essential for a botanical institution), he would do so on his own schedule. First he wanted to erect walls and dig trenches—it wouldn't be a garden without living plants. "The museum and library, must be built after the plant houses," he informed Engelmann. Even the ornamental plantings should come before the scientifically useful ones. He assured Engelmann that he was not acting idly, but that if he did all that Engelmann wanted him to, he "should make a fatigue of a pleasure."[69] And pleasure for Shaw, in building and sharing his garden, was important.

Despite his frustrations with Shaw, Engelmann came to realize that this wealthy amateur gardener might accomplish what the old Western Academy had failed to do—establish an important botanical institution near the city. So in the spring of 1856, already overburdened with responsibilities, Engelmann agreed to act as Shaw's agent abroad. He was soon to travel to Europe anyway for an extended visit with relatives and friends, and it would fit in well with his plans to represent Shaw's interests there—to meet with Hooker, visit botanical gardens, gather ideas, perhaps even seeds, for the St. Louis project. Furthermore, Engelmann could use his connections to search for books and plant specimens for Shaw. Engelmann asked in return only that Shaw take care of some germinating cactus seeds while he was away.

The trip yielded results. Shaw had Engelmann send him as many plans and catalogues of various gardens as he could find. He also ordered books from a list suggested by Hooker—just three books at first, to help him learn more botany and horticulture, but he eventually authorized Engelmann to buy thirty-four volumes in Europe and to spend up to $100 for their purchase.[70] The greatest treasure, however, came when Engelmann learned from his European contacts that the herbarium of the late German professor Johann Jakob Bernhardi, containing an estimated thirty thousand plant species, was for sale for $1,000. Shaw authorized Engelmann to negotiate the purchase, which he did, getting the price lowered finally to $600. It was an incredible bargain. Shaw now owned

---

68. Engelmann to Benjamin A. Soulard, November 24, 1854, Soulard Papers, Missouri Historical Society Archives.
69. Shaw to Engelmann, September 15, 1857, Engelmann Papers.
70. Shaw to Engelmann, October 18, 1856; January 13, 1858, Engelmann Papers.

Shaw's Museum (from stereoscopic photograph by Boehl and Koenig, 1868–1874). Shaw convinced his advisor Engelmann that he should build his museum only after having created the garden itself. *Missouri Historical Society*

a vast collection of plants from Europe and Asia, many of them type specimens (used to establish the original scientific description of the species). No other single collection in the United States rivaled it.[71]

After his return to America, Engelmann advised Shaw in a matter that set the tone of the institution. Shaw was puzzled by what to name his garden. He thought seriously about placing the words "Hort. Bot. Missouriensis" over the entrance, no doubt believing that Latin would lend scholarly prestige to his endeavor. But Engelmann, who had named his share of plants in Latin, thought otherwise. Why not keep things simple and understandable, he suggested, and use the English equivalent? Shaw agreed, and named his establishment the Missouri Botanical Garden.[72]

Progress and Disaster

Besides acting as Shaw's agent abroad, Engelmann looked after numerous other scientific matters. He crisscrossed Europe several times in two years, visiting cactus growers in Germany, England, and France, supervising artists in Paris who were engraving illustrations for his cactus report, and informing "scientific men and societies" everywhere about the new academy in St. Louis.[73] He no doubt impressed colleagues in London, Paris, Geneva, Naples, Rome, Vienna, Leipzig, Frankfurt, and Berlin with descriptions of Hayden's fossils and the plans of the academy shortly to begin publishing its transactions. He hardly would have missed the opportunity to mention the new society to Alexander von Humboldt—who had explored in South America and dined with Thomas Jefferson—when he visited the aged naturalist in Berlin in 1857.[74]

The academy, meanwhile, published the first number of its *Transactions* in 1856, running off 1,000 copies for distribution. It earmarked 150 copies for the Smithsonian Institution to spread overseas; another 25 copies for Engelmann to give to private individuals in Europe; and 49 copies to send to scientific academies and medical journals in the United States, from Boston and Philadelphia to New Orleans and California. The response was quick and encouraging. Societies began sending copies of their proceedings and transactions as well as

---

71. Georg Mettenius to Engelmann, October 31, 1857, Engelmann Papers. See also William D'Arcy, "Mysteries and Treasures in Bernhardi's Herbarium," *Missouri Botanical Garden Bulletin* 59:1 (January–February 1971): 20–25.
72. Engelmann to Asa Gray, September 28, 1858, GHAH.
73. Minutes of council meetings of the Academy of Science, September 6, 1858.
74. Engelmann to William Emory, July 8, 1857, draft in Engelmann Papers.

Missouri Botanical Garden entryway (from stereoscopic photograph by Robert Benecke, c. 1875). It was Engelmann who convinced Shaw to inscribe the garden's simple English-language title—rather than its more obscure Latin equivalent—across the public entrance, where it remains today. *Missouri Historical Society*

other scientific works in exchange, and the academy almost immediately authorized an additional 100 copies for distribution, including 10 copies to circulate, via Charles Chouteau, throughout "Indian country."[75]

The academy at last had entered the larger world of scientific publication. No longer would local scientists be forced to submit papers to distant journals to spread their ideas and receive recognition. By publishing first in the *Transactions*, scientists with international reputations like Engelmann increased the exchange value of each issue. Thus even though the academy struggled to publish each year, spending most of the membership dues on that activity alone, the result was worthwhile. By 1860 the academy had issued the fourth number of its *Transactions*, sending copies to 226 "societies, universities, and authors" in the United States and Europe.[76] Its library grew each year from exchanges, receiving publications that members could not have afforded to buy outright.

Even the Civil War did not greatly affect this exchange system. During the years of conflict the academy failed to issue its *Transactions* only once, in 1862, and the fault lay not with the war but with the academy's treasury, which shrank too small to allow publication. Engelmann, by then the president, appealed to the members for more generosity, and the following year the *Transactions* appeared once again on schedule. Of course, war could have greatly devastated science in the city at any time, as the members were well aware. Engelmann, for instance, feared that if opposing armies were to meet near St. Louis, a natural battlefield would be the Missouri Botanical Garden, with its "substantial buildings," long walls, and high ground.[77] Fortunately such a battle never occurred, but war did leave its mark on the academy.

In the early days of the conflict the surgeon Joseph Nash McDowell, head of McDowell Medical College and a Southern sympathizer, abandoned his institution to Union forces, who confiscated the building and turned it into the Gratiot Street prison for Confederate soldiers. The Western Sanitary Commission in St. Louis asked the Academy of Science in 1862 to "receive on deposit the remains of the McDowell Collection of Natural History" still at the prison. The academy agreed to take "the best preserved anatomical preparations, specimens of animals, minerals, and other articles of value . . . in trust for safe keeping."[78]

The academy kept them safe throughout the rest of the war and for several years beyond. Then, in 1869, disaster struck. As the academy was preparing to move its library and museum to larger quarters at the new Polytechnic Building

---

75. Minutes of council meetings of the Academy of Science, March 23, 1857; April 6, 1857; April 27, 1857.
76. Journal of Proceedings, annual address, January 7, 1861, in *Transactions* 2: 145.
77. Engelmann to Asa Gray, August 17, 1861, GHAH.
78. *Transactions* 2: 192.

downtown, fire swept through the academy's room at Pope's Medical College. Quick action saved most of the books and publications, but flames consumed nearly all the specimens. Gone were the animal skins, the fossils, the geological material from William Clark; the collections made by Engelmann and the Western Academy; the fossils that Hayden brought back from Nebraska Territory; the donations of Chouteau, Koch, and others—all lost. Had these things survived, they surely would have merited considerable scientific and historical attention. Instead, we can only imagine them.[79]

The Center Shifts

The academy struggled for years to recover. Though the members finally moved into the Polytechnic Building (which also housed the Missouri Historical Society), the rent-free rooms were inadequate for library use and for displaying what few specimens they had. Members also were too depressed by their loss and the aftermath of the Civil War to rebuild their collection. Besides, the general public seemed unsupportive. In 1872 the new president of the academy, James B. Eads, accused Missourians of being too ignorant of the benefits of science to allow their state legislature to help the academy, even though "no institution in Missouri more richly deserves the patronage of the State."[80] Missourians might have felt they were benefiting enough already by supporting a state entomologist, Charles V. Riley, who in the 1870s provided them with practical, informative, and well-illustrated annual reports that won even the admiration of Charles Darwin. Perhaps the public also viewed the academy as a bit old-fashioned, still collecting and describing in pre-Darwinian ways—efforts that had paid off in the antebellum years, but which now seemed out of step with the times.

St. Louis had changed greatly since the small group of young men interested in science first gathered there to form a scientific society. The city in the Gilded Age no longer enjoyed the advantage of being the great hub of commercial and scientific enterprise created by river traffic. Instead, Chicago had surpassed St. Louis in the commercial race to connect east and west by rail. Specimens gathered in western lands could bypass St. Louis all together. Nor was the small-business model that Engelmann developed for scientific collecting in the 1840s

---

79. Journal of Proceedings, January 3, 1882, in *Transactions* 4: lxix; Mary J. Klem, "The History of Science in St. Louis," *Transactions* 22: 122.

80. James B. Eads, *Inaugural Address to the Academy of Science by the President* (St. Louis: n.p., 1872).

and '50s still up to the task. Like the growth of big industries and monopolies in post–Civil War America, science now required large amounts of capital and specialized labor for its endeavors—more than a single academy of science could provide.

Federal agencies and various universities around the country became the new engines of research. The government, for instance, directed many western expeditions, everything from railroad surveys to a census of U.S. forests. Universities supported expensive research facilities, like astronomical observatories, which the Academy of Science could not afford. And universities played an even more important role in changing the ways of research, because they trained and employed the next generation of scientific specialists, who no longer felt obliged to become physicians or engineers in order to pursue their science.[81] The able amateur at last yielded to the full-time, wage-earning scientist.

For a while it seemed as if the Academy of Science of St. Louis might benefit from these changes. In the 1880s the organization moved to new quarters at Washington University, thus connecting itself, at least physically, with a strong ally. George Engelmann was once again president of the academy, and he spoke with satisfaction of "ample room, surrounded by our rapidly increasing library and by the rudiments of our museum." But his vision of the future was really a reflection of the past. To him the academy was now "firmly established as a central point of Science in the Mississippi valley." Yet even as he proclaimed this he acknowledged his uneasiness with the new location; the campus bustled around him with scientific activity, while "we work along in our unpretentious way."[82] Though the academy continued for years to serve as a forum for scientific discussion, even including the general public at its meetings, it eventually moved from Washington University and ceased to be the "central point" of research in the city.[83]

Henry Shaw, who was still alive when Engelmann died in 1884, understood the importance of the university over the academy. Shaw had been a longtime member of the Academy of Science, but never a large benefactor of the institution, even after the death of his friend who had worked so hard to keep the academy alive. Shaw decided on another way to honor Engelmann, one that would be more lasting. In 1885, a year after Engelmann's death, Shaw endowed

---

81. Hendrickson, "Science and Culture," 339–40. See also Howard S. Miller, *Dollars for Research: Science and Its Patrons in Nineteenth-Century America* (Seattle: University of Washington Press, 1970), and Robert V. Bruce, *The Launching of Modern American Science, 1846–1876* (New York: Alfred A. Knopf, 1987).

82. President's address, January 2, 1883, in *Transactions* 4: lxxxii.

83. For more on the later years of the academy, see John R. Hensley, "The Academy of Science of St. Louis, 1856–1988," *Transactions* 33 (1988).

a School of Botany with an Engelmann Professorship at Washington University. Thus he forged a link between the Missouri Botanical Garden, which Engelmann also had helped to found, and the university that would help keep the Garden at the forefront of botanical research for years to come. It was a fitting tribute to Engelmann, who, along with many others filled with entrepreneurial spirit and zeal, had first brought science to St. Louis.

Public Education in Nineteenth-Century St. Louis

William J. Reese

The nineteenth century was a great age of institution building: from prisons and workhouses to hospitals and asylums. For a variety of reasons, Americans turned increasingly to the power of municipalities and the state to cure the ill, punish the fallen, and educate the young. And no institution held greater hope among the citizenry for the nation's welfare and future than its growing system of public schools. Historically, schools in Europe had been controlled by the established church, were rarely free, and were usually segregated along the lines of religion, gender, and social class. But in America, both in countryside and city, free public schools became commonplace by the middle of the nineteenth century and provided the setting in which most children (at least in the northern states) received their basic formal education. In hyperbole common to the times, the nation's most famous educational reformer, Horace Mann of Massachusetts, claimed at midcentury that free public schools were the greatest invention of the age. Along with educators and reform-minded citizens across the northern states, he claimed that schools would reduce illiteracy and social strife by teaching basic knowledge, common moral beliefs, and the values needed to live harmoniously in the American republic. They would also acculturate immigrants and even reduce if not eliminate poverty by teaching the work ethic and personal responsibility to every child.[1]

Americans today routinely condemn the quality of public schools in urban areas. By the late nineteenth century, however, urban schools were often regarded as the most progressive in the nation. Many citizens, including most educational leaders, routinely pointed to their innovative features: the establishment of graded classrooms, the hiring of women in the elementary grades,

---

1. The standard history of the common schools is Carl F. Kaestle's *Pillars of the Republic: Common Schools and American Society, 1780–1860* (New York: Hill & Wang, 1983), which provides citations on the large secondary literature on public education. On Mann, see Jonathan Messerli, *Horace Mann: A Biography* (New York: Alfred A. Knopf, 1972).

the broadened curriculum, the creation of free high schools, and the appointment of administrators to implement new programs and policies and to make the system more publicly accountable. The one- and two-room schools found in most rural districts—so-called "little red schoolhouses"—have become part of a collective nostalgia for all things small and beautiful. At the time, however, rural schools were frequently regarded as inferior, offering a meager curriculum taught by underprepared, poorly paid teachers. The best, most ambitious rural teachers often sought positions in the cities. Indeed, unlike citizens today, many nineteenth-century Americans saw urban schools as in the vanguard of educational progress. And few cities played a more conspicuous role nationally in the rise of the public schools than St. Louis during the last half of the nineteenth century.[2]

From inauspicious beginnings, the St. Louis public schools grew dramatically during the long St. Louis residence of Henry Shaw, becoming nationally prominent after the Civil War. While only a handful of pupils entered the city's first public schools when they opened in the 1830s, over eighty thousand young people did so by the turn of the twentieth century. By 1900, about 82 percent of all children were enrolled in the public system as opposed to private schools; most children attended classes at least through the grammar grades. High school enrollments also grew spectacularly in the new century.[3] Growth is not synonymous with goodness or progress, but it was how many educators viewed their world. As early as 1858, Ira Divoll, the superintendent of schools, boasted that "the St. Louis Public Schools, though only in their infancy, far outnumber all the other schools of the city, and it is believed that the instruction given in them is such as commends itself to all classes of reflecting parents, and that it promotes, in no small degree, practical virtue and morality among the people."[4] Reflecting a widespread faith that the public schools could promote the common welfare, another administrator similarly affirmed decades later that "[t]he great social mission of the public schools is to unite all classes of society in their rooms in the common educational preparation for life. The children of all classes, of all social ranks, of all shades of belief, affiliate during the . . . years of their school life, form ties of friendship and affection, and cultivate mutual regard and good will." Echoing sentiments that educators and statesmen

---

2. Carl F. Kaestle, "Rural Schools in the Early Republic" and "Urban Education and the Expansion of Charity Schooling," chaps. 2 and 3 in *Pillars of the Republic;* William J. Reese, *The Origins of the American High School* (New Haven: Yale University Press, 1995), 21–27.

3. *Forty-sixth Annual Report of the Board of Education of the City of St. Louis, Mo., for the Year Ending June 30, 1900* (St. Louis: Buxton & Skinner Stationery Co., 1901), 47–48, 110.

4. *Fourth Annual Report of the President, Superintendent and Secretary, to the Board of St. Louis Public Schools, for the Year Ending July 1, 1858* (St. Louis: R. P. Studley, 1858), 47.

shared throughout the century, he concluded that public schools were essential "for the perpetuity of free institutions" and the preservation of the "social and industrial order."[5]

Similar assertions were commonly registered in St. Louis and in urban systems across the nation. The schools aimed to ensure the stability and improvement of the republic, to lessen crime and promote morality, and to reduce the tensions between rich and poor, native-born and immigrant. During Reconstruction in the 1870s, Radical Republicans even lobbied for more opportunities and social justice for African Americans.[6] The gap between rhetoric and social practice nevertheless often remained wide. Like many public institutions, schools often promised more than they could deliver, even if St. Louis citizens supported them with their tax dollars and, generally, endorsed them over private schools. Acrimonious debates about educational policy—that the schools taught too many subjects or too few, emphasized the basics too strenuously or not enough, should teach foreign languages or only English, or cost too much or were too penurious—periodically surfaced throughout the period. Support for the public schools, however, remained relatively strong even through hard economic times, which included the depression of the 1870s, a severe turndown following recovery in the 1880s, and the onset of the nation's worst depression to date in 1893. Despite local disagreements about particular educational policies and uncertain economic conditions, the local schools often found themselves in the national spotlight.[7]

Explaining why certain educational leaders in St. Louis became national figures defies easy explanation. It resulted from a combination of chance, an ability to act creatively in propitious moments, and popular interest nationally in educational experiments under way in America's cities. Whatever the explanation, the St. Louis schools and the ideas of locally prominent educators were widely discussed in educational meetings, the popular press, and professional journals around the country. Most attention focused on the writings and activities of three important figures: William T. Harris, the superintendent between 1868 and 1880, who emerged as one of America's most prominent educational leaders and intellectuals; Susan Blow, who with Harris's encouragement in the 1870s helped establish the nation's first extensive system of public kindergartens; and Calvin M. Woodward, who from his post at Washington University and on the school board fought for various manual training and practical courses to better

---

5. *Forty-sixth Annual Report*, 153.
6. Elinor Mondale Gersman, "The Development of Public Education for Blacks in Nineteenth-Century St. Louis," *Journal of Negro Education* 41 (Winter 1972): 35–47.
7. The most important history of the St. Louis schools is Selwyn K. Troen, *The Public and the Schools: Shaping the St. Louis System, 1838–1920* (Columbia: University of Missouri Press, 1975).

prepare young people for the workplace. Examining their key ideas in the context of their city and their times helps illuminate the nature and character of public education after the Civil War, when attending school became a familiar experience in the lives of children.

The rise of William T. Harris is basic to the evolving history of the schools. Harris was a Connecticut Yankee, born to Calvinist parents in 1835. After attending various lower schools in the countryside and city, he enrolled at Yale College. There he lost his religious faith, despite the school's religious orthodoxy, after flirting with—among other attractions—mesmerism, the water cure, and transcendentalism. He withdrew from Yale in his junior year, heading west like many young people searching for their future. Soon the bookish Harris became one of the nation's leading exponents of German idealism. He helped form what became known as the St. Louis school of philosophy and translated Hegel's *Logic* into English. Few cities at the time could claim a philosopher-king as their school superintendent. But the enterprising Harris had quickly risen up the ranks from teacher to principal to superintendent. His annual reports to the school board, issued between 1869 and 1880, were lengthy, often translated into German (given the prominence of German immigrants in the city), and distributed to educational leaders across the nation. Admiring of centralized authority, he nevertheless provided considerable autonomy to local principals of the burgeoning school system and became famous as a system builder and especially as a defender of the humanistic purposes of public education. Harris would serve on virtually every national blue-ribbon committee that dealt with educational policy in the late nineteenth century. After departing St. Louis, he also gained an unusual bully pulpit when he became the U.S. Commissioner of Education: he served between 1889 and 1906 under both Republican and Democratic administrations, the longest tenure of any person in that office.[8]

The post–Civil War years were exciting ones in the St. Louis schools. By the time Harris became superintendent in 1868, some of the basic organizational features of a fledgling bureaucracy were in place. Harris would add his own unique contributions to the schools and their organization, but much had been accomplished before his tenure. From the 1850s onward, administrators labored to provide more uniform, standard amounts of learning to the young by creating more age-graded classrooms, mostly taught on the elementary level by

---

8. On Harris, see Kurt F. Leidecker, *Yankee Teacher: The Life of William T. Harris* (New York: Philosophical Library, 1946); Neil McCluskey, *Public Schools and Moral Education: The Influence of Horace Mann, William Torrey Harris, and John Dewey* (New York: Columbia University Press, 1958); Herbert M. Kliebard, *The Struggle for the American Curriculum, 1893–1958* (New York: Routledge, 1995), 30–35; and William J. Reese, "The Philosopher-King of St. Louis," chap. 7 in *Curriculum and Consequence: Herbert M. Kliebard and the Promise of Schooling*, ed. Barry M. Franklin (New York: Teachers College, Columbia University, 2000).

William Torrey Harris (steel engraving by J. C. Buttre after a photograph by Scholten, n.d.). Harris, the "philosopher-king" of St. Louis's public school system, dominated both the local and national educational scene with his writings and efforts on behalf of a free liberal-arts education in the late 1800s. *Missouri Historical Society*

young women. The vast majority of children were concentrated in the elementary grades, where overcrowding was common, as in urban systems everywhere in the coming decades. In 1866, Superintendent Divoll emphasized the superiority of graded classes, in contrast to the typical ungraded rural school. "The advantages which will accrue from this classification of studies are obvious," he wrote. "The teacher or parent can tell the scholar's proficiency in all his studies by knowing his advancement in any one."[9] And the achievements of children of

9. *Eleventh Annual Report of the Board of Directors of the St. Louis Public Schools, for the Year Ending August 1, 1866* (St. Louis: R. P. Studley & Co., 1866), 43.

the same age in the same grade in different schools could be compared, indicating points of pride or shame. Perfectly age-graded classrooms were the ideal of every school administrator for the remainder of the century.

School enrollments had doubled in the 1870s to about fifty thousand pupils when Harris retired as superintendent. Between 1867 and 1881, the system increased from 30 to 103 schoolhouses, and the newest buildings (all larger than earlier ones) typically had eighteen rooms. Coeducation became the norm, in contrast with many private schools, and this helped reduce costs. Among the 103 schools was a central high school, opened in 1853, and a normal school which appeared soon after; the latter trained the vast majority of local elementary teachers.[10] St. Louis High enrolled only a tiny percentage of the total student body. But it was called the "people's college" for providing free secondary education for talented pupils, who had to pass an examination to gain entry. Featured in national publications such as the *American Journal of Education*, St. Louis High offered scholars a high-quality academic education and helped lure middle-class families into the system, away from competing private schools. In addition, German language instruction, championed by immigrant partisans on the board of education, was extensive in the school system, though always controversial. Finally, the teaching force dramatically increased between 1867 and 1881, from 220 to nearly 1,000. The teachers' labors were intensive. Except for the high school, where smaller classes prevailed, the school board mandated a minimum of fifty-eight pupils per classroom, a reflection of the inability of schools to keep up with rising enrollments.[11]

Large class sizes in the elementary- and grammar-level grades meant that St. Louis's schools retained their traditional reliance upon textbooks, rote memorization, and recitation, prevalent features of the typical classroom. Harris and others insisted that textbooks had a democratizing influence, providing each child in theory with access to the same knowledge. Schoolmen realized that they could not build schools or classrooms fast enough to meet the popular demand, or hire a sufficient number of teachers to reduce class size. Given the realities of large classes, textbooks seemed the fairest and most efficient way to educate the young. John Tice, the superintendent of schools in 1854, nevertheless typically complained that an overreliance on textbooks had some baleful effects. Requiring pupils to memorize vast quantities of material was a time-honored practice; only through daily exercise could the mind (like other muscles, it was believed) gain strength. But improving the mind sometimes led to inhumane pedagogy.

---

10. *Twenty-seventh Annual Report of the Board of President and Directors of the St. Louis Public Schools, for the Year Ending August 1, 1881* (St. Louis: Slawson & Co., 1882), 36, 39; and Troen, *Public and the Schools*, 15–17, 22–27.

11. *Eleventh Annual Report*, 38, 42; and Reese, *Origins of the American High School*, 90.

Like many educators who followed him, Tice believed that pupils spent too much time "committing to memory whole volumes of abstract facts," as well as the "barbarous names of villages which they will never hear of after they are done reciting their lessons!"[12]

What was true of geography was true of the basic subjects: reading, writing, and arithmetic, as well as history. Nineteenth-century educators throughout the nation placed an enormous emphasis on memorization and oral recitation. Pupils memorized the names of mountains and rivers, the rules of grammar, many facts of American history, the multiplication table, patriotic speeches, song lyrics, and much more. The goals of education were expansive, embracing moral as well as mental discipline. Teachers wanted all children to learn the basic branches of knowledge, Christian morality, punctuality, and deference to authority, but also some sense of the grandeur of American history, the nation's material wealth, and its superiority to Europe, which the children's textbooks said were often ruled by evil monarchs, the papacy, and aristocrats.[13]

When Harris became superintendent in 1868, he therefore inherited a complex enterprise, one that grew larger and more complicated throughout his tenure. Along with other public school leaders, Harris believed that individuals in a modern society could not rely on the traditional institutions of family and church alone to prepare children for the future. An urban industrial society needed schools to teach everyone common values, precepts, and knowledge; otherwise, Harris predicted, there would be social chaos and disintegration. Like Thomas Jefferson earlier and many contemporary educators, Harris wanted schools to play an important role in identifying and rewarding individual talent. This would promote social mobility and thus help keep America's social order fluid. Public schools also had a moral and civic obligation to teach all children, rich and poor, native-born and immigrant, the values of a common culture. More liberal on racial matters than the times, Harris, a Republican and Union supporter, also favored racial integration, which had few champions beyond a handful of Radical Republicans in the 1870s. In an ideal school system, which he knew did not exist, all children would have access to the same resources, quality teachers, and overall opportunities. Only then, in Hegelian fashion, could the tension between the individual and society resolve itself into a synthesis of social harmony, justice, and industrial progress. By attending school together, children would learn a "common stock of ideas" and

---

12. *First Annual Report of the General Superintendent of the St. Louis Public Schools, for the Year Ending July 1, 1854* (St. Louis: Printed at the Republican Office, 1854), 13.

13. These multiple goals for the schools, described in the *First Annual Report*, were commonly invoked in the board's reports throughout the remainder of the century. Complaints about the dull teaching in the schools were ubiquitous.

values, thereby reducing social tensions and strengthening the nation's civic culture.[14]

To Harris and most teachers, the moral aims of education were unambiguous: children were expected to be honest, hard-working, punctual, and virtuous. These time-tested values seemed particularly appropriate, he thought, in an industrial age, where showing up on time was basic to economic survival. School prayers and Bible reading had long been banned in the St. Louis schools, a policy Harris defended despite recurrent sectarian attacks on its "godless" implications. Harris was especially adamant in his defense of the humanistic and intellectual purposes of public education. In somewhat mystical language, he said the schools were duty-bound to open up what he called "the five windows of the soul" for each child. Said plainly, he promoted an academic, humanistic education for everyone. He believed that all children should be exposed to the canon of Western thought, widening their intellectual horizons beyond the more limited vistas of family, church, and neighborhood. All pupils should, according to this view, master five broad domains of knowledge: arithmetic and mathematics, which taught unique ways of seeing reality; geography, which extended one's visual and mental perspectives; history, which showed what humans were capable of; grammar, which allowed full expression of one's views; and the arts and literature, which provided multiple aesthetic riches.[15]

Harris was a cosmopolitan intellectual but also an effective administrator and politician. The *St. Louis Post-Dispatch* perceptively remarked that "Mr. Harris is a transcendental philosopher, and when he gets hold of a Philosophy of the Conditioned he can puzzle a spelling class; but when he takes hold of a plain question of fact, or explains the management of the public schools, he can satisfy the dullest intellect that his dealings with the abstruse mysteries of Kant and Hegel have not unfitted him for his practical work as a Superintendent."[16] Harris admired German culture, especially its intellectual contributions in literature, the arts, and the sciences. He worked well with German-Americans on the school board. With them he endorsed the teaching of German in the schools, which helped popularize the system, since this lured many immigrant children out of private academies and religious schools. Harris also pioneered in the creation of science instruction in the elementary grades and of more rapid promotion policies; both ideas were decades ahead of the times. Because he

---

14. *Fifteenth Annual Report of the Board of Directors of the St. Louis Public Schools, for the Year Ending August 1, 1869* (St. Louis: Missouri Democrat Book and Job Printing House, 1870), 110. For a more complete assessment of Harris, see Reese, "Philosopher-King of St. Louis."
15. Kliebard, *Struggle for the American Curriculum*, 15, 32–34, 55–56.
16. *St. Louis Globe-Democrat,* May 8, 1975; quoted in Leidecker, *Yankee Teacher,* 262.

was identified nationally as a champion of the humanities, however, he became known as a conservative: he endorsed academics above all, and ridiculed the overly romantic, sentimental views of children embraced by some educational reformers. A formidable platform debater, he also sneered at those who wanted less academics and more vocational subjects in the schools, and they returned the favor.[17]

Harris defended traditional academics but not poor, lifeless teaching. Long after he left St. Louis, he continued to criticize the mind-numbing pedagogical practices found in many schools. Yet Harris realized why traditional practices died very slowly. He knew that the pool of outstanding teachers at any given moment was never large. Moreover, they taught very large classes and also faced parents who usually wanted the basics taught in familiar ways. And so teachers understandably relied heavily on custom: they assigned homework in ubiquitous textbooks, their pupils tried to recite accurately what they learned, and both then moved on to the next lesson. While textbooks had their limitations, Harris often defended them from romantics who wanted to dispense with them; in theory, textbooks offered all children, irrespective of social background, access to the same knowledge. And, given the number of uninspiring teachers found in many schools, textbooks offered something valuable and authoritative for everyone. "The printed page," Harris wrote in 1870, "is the mighty Aladdin's lamp, which gives to the meanest citizen the power to lay a spell on time and space. It is the book alone that is reliable for exhaustive information."[18] And every small step taken in mastering knowledge moved the child closer to intellectual maturity and reinforced the habit of learning on his or her own.

Critics throughout the century, including Harris, realized that an overreliance on textbooks often led to mediocre instruction. This was an age of educational reform, full of schemes for human improvement, including more humane and effective pedagogy. Like other school leaders, Harris studied what advanced thinkers in Europe were saying about children and how they best learned. While he criticized many of the assumptions of child-centered education that arose from romanticism, he joined with others in a movement that ultimately led to the adoption of free public kindergartens in St. Louis in the early 1870s. It was the first major school system to do so on a large scale. The kindergartens ultimately faced their own political opposition and did not lead

---

17. Troen, *Public and the Schools*, 61–65. Harris's conservatism was the theme of a leading critic, Merle Curti, *The Social Ideas of American Educators* (Totowa, N.J.: Littlefield, Adams, [1935?]).

18. *Fifteenth Annual Report*, 27.

Susan Blow (photograph by Emil Boehl of pastel by C. F. Maury). St. Louis educator Blow, a follower of Harris, introduced German idealist concepts of early-childhood learning into the American classroom. *Missouri Historical Society*

to the transformation of pedagogy in the rest of the system. But the experiment attracted considerable national and even international attention.[19]

Very familiar with the child-centered educational experiments emerging on the continent, Harris encouraged a young associate named Susan Blow to study kindergarten methods. Together they helped gather support in the community and on the school board for early-childhood education. Born in St. Louis in 1843 to a rich and influential family, Blow was a deeply religious person, educated by private tutors and at an elite eastern academy. She traveled abroad, examining kindergartens firsthand and studying the often abstruse writings of Friedrich Froebel, the German inventor of the kindergarten.[20] The kindergarten was the leading romantic reform among contemporary educators. To many it promised a more relaxed and nonbookish environment for children, where sympathetic female teachers would employ gentle approaches in teaching the very young. Blow became the key figure in the kindergarten movement in St. Louis in the 1870s. She believed, according to historian Barbara Beatty, that Froebel "had unlocked an ancient and secret code" about children and how they best learned. With Harris's help, Blow also published translations of Froebel's songs and music. She was a dynamic public speaker and became a well-known author. By the turn of the century, she was a prominent defender of orthodox kindergarten practices, insisting that early-childhood educators should not deviate from Froebel's original ideas.[21]

First serving as a substitute teacher, Blow ultimately supervised a large network of volunteer and then salaried kindergarten instructors in St. Louis. From a modest experiment in 1873, when the first local kindergarten opened, the innovation soon spread across the city, despite some noisy opposition to the German-inspired reform on the grounds of ideology and expense. Blow and Harris countered that opposition with a pragmatic approach that contrasted sharply with what they believed to be the naive and sentimental views of those kindergarten advocates who claimed to have created veritable paradises for children. Both of these educators criticized these romantics and believed that taxpayers would more likely favor kindergartens as a way of teaching the very

---

19. "The Beginning of the Public School Kindergarten Movement," chap. 5 in Troen, *Public and the Schools*.
20. Barbara Beatty, *Preschool Education in America: The Culture of Young Children from the Colonial Era to the Present* (New Haven: Yale University Press, 1995), 52–53, 64–67; Barbara Beatty, "Susan Elizabeth Blow," in *Historical Dictionary of American Education*, ed. Richard J. Altenbaugh (Westport, Conn.: Greenwood Press, 1999), 48–49. For Harris's views on progressive writers on education, see Reese, "Philosopher-King," 170–75.
21. Beatty, *Preschool Education in America*, 46, 71.

Interior of Shepard Kindergarten (collotype by Robert Benecke, c. 1885). The Shepard School incorporated the concepts of kindergarten learning into its design. *Missouri Historical Society*

young in innovative ways while preparing them for the more structured elementary grades. Kindergartens would also serve as a bridge between the informality of the home and the more formal disciplinary and learning environment of the school.[22]

Like European child-centered educators, Blow and Harris realized that the young learned through sensory experiences, not books alone. Children in kindergartens engaged in a series of structured and graduated pedagogical exercises that Froebel had called "gifts" and "occupations."[23] These lessons, however adapted in different classrooms, everywhere aimed to teach social cooperation, manual dexterity, bodily control, and numerous abstract concepts about space, form, and the mathematical principles that helped structure and unify the physical world. The St. Louis kindergartens did not promote unstructured play but focused on the skills and values young children needed for their intellectual and moral development. The "child's garden" was in reality a bit of a workshop, where children were actively involved in learning, gaining the discipline needed in the elementary grades.

22. Ibid., 65–67. See also the lengthy commentaries by Harris and Blow in the annual reports to the school board in the 1870s.
23. There is a voluminous literature on Froebel and his ideas; a good introduction to the subject is Norman Brosterman, *Inventing Kindergarten* (New York: Harry N. Abrams, 1997), chap. 1.

As a result of Blow's influence, the local kindergartens grew increasingly popular and soon gained national recognition. She trained and supervised hundreds of teachers and inspired countless others through her lectures and published writings. Both Blow and Harris wanted to make the kindergartens available for free to all children, though they thought the poor especially needed access to them. In the early 1870s, Harris investigated the social conditions of different neighborhoods in St. Louis, noting the desperate situation of the poor, especially along the levee. He found poignant examples of personal misery and social despair that programs such as kindergartens might ameliorate. Like Harris, her staunch ally, Blow believed that the kindergartens might help the poor overcome some of their social disadvantages by teaching them discipline and self-control and by offering an alternative to the influence of the streets and the slum.[24]

Throughout the 1870s, critics of the kindergartens continued to question their legitimacy. They were called expensive frills that absorbed money better spent elsewhere in the schools. In his last year as superintendent, Harris nonetheless insisted that every child could benefit by attending them. "If he is a child of poverty, he is saved by the good associations and the industrial and intellectual training that he gets. If he is a child of wealth, he is saved by the kindergarten from ruin through self-indulgence and the corruption ensuing on weak management in the family." The interaction of rich and poor would thus nurture a more socially harmonious society. By 1880, even the most ill-treated group in St. Louis, African Americans, had gained access to the innovation, albeit in segregated schools.[25] But legal challenges to the existence of free public kindergartens soon intensified.

The free-kindergarten experiment ended in 1883 when the courts and the state legislature forbade the use of public funds to educate children under the age of six. The kindergartens were reestablished on a tuition basis, and the city remained well-known for its faith in the power of early-childhood education despite this legal setback. Due to ill health, Blow left St. Louis in 1884 for the East Coast, where she continued to champion Froebel's philosophy. Harris's successors as superintendent, while forced to charge fees, nevertheless continued to champion the kindergartens, whose relaxed and pleasant yet structured

---

24. See Harris's comments in the *Seventeenth Annual Report of the Board of Directors of the St. Louis Public Schools, for the Year Ending August 1, 1871* (St. Louis: Plate, Olshausen & Co., 1872), 37–38, and in the *Nineteenth Annual Report of the Board of Directors of the St. Louis Public Schools, for the Year Ending August 1, 1873* (St. Louis: Democrat Lithographic and Printing Company, 1874), 18–19.

25. *Twenty-fifth Annual Report of the Board of Directors of the St. Louis Public Schools, for the Year Ending August 1, 1879* (St. Louis: G.I. Jones & Co., 1880), 136; Troen, *Public and the Schools*, 108.

ambience, he felt, deserved imitation throughout the elementary grades.²⁶ Superintendent Edward H. Long (1880–1895), a firm advocate of early-childhood education, insisted in 1894 that "[i]f Froebel's methods were fully understood by all teachers of the primary and intermediate grades and were applied in giving instruction in the branches taught in these grades much more satisfactory results would be secured."²⁷

Long's successor, F. Louis Soldan, even claimed in 1897 that the lower grades were becoming transformed by kindergarten methods. Over ten thousand children attended public kindergartens annually by the close of the century, and approaches to teaching, Soldan said, had "brought about an education revolution by driving out the old formalism and mechanical text-book study." He nevertheless conceded in the next breath that tradition generally ruled in most elementary classrooms. Something less than a revolution had actually occurred. As in the 1850s, children still studied the basic subjects, memorized material from textbooks, and recited what they knew to teachers. "Study and recitation," Soldan reluctantly admitted, "are the chief activities of a child's school life."²⁸

While the kindergarten had failed to transform school practice in the elementary grades, the issues raised by romantic critics had nevertheless left their mark on educational thought, and ultimately led to some important curricular changes despite considerable resistance. The complaint among students that schools were often boring and unappealing was, of course, hardly news by the 1890s. Romantics and nonromantics alike since midcentury knew that schools were not a child's paradise and most teachers and administrators continued to emphasize the importance of work, not play, and study, not sport. No one defended the Gradgrinds in their midst, and a succession of superintendents and other citizens in the last half of the nineteenth century complained about boring classes, dull teaching, and the resultant apathy of some teachers and many pupils. But it remained easier to complain about than eliminate the most reviled classroom practices.

The kindergarten, however, contributed notably to the debate over how to enliven instruction and make it more meaningful to children. It promised what many reformers at the time called a "new education," one that tapped children's senses, ended their passivity, and sought ways to instruct beyond the usual trio of reading, memorizing, and reciting. Children in the kindergarten worked with clay, paints, and paper, played with differently sized objects, and

26. Troen, *Public and the Schools*, 112–14.
27. *Thirty-ninth Annual Report of the Board of President and Directors of the St. Louis Public Schools, for the Year Ending June 30, 1893* (St. Louis: Nixon-Jones Printing Co., 1894), 89.
28. *Forty-second Annual Report of the Board of President and Directors of the St. Louis Public Schools, for the Year Ending June 30, 1896* (St. Louis: Nixon-Jones Printing Co., 1897), 98, 111.

were encouraged to cooperate and interact with each other more than was true in the higher grades. Most important, some educators, whether or not they had read or even understood Froebel's writings, felt encouraged to more openly question whether books should predominate in the schools. They wanted to make children more active in the learning process and teachers more sensitive to training their pupils' hands and bodies as well as their minds. In addition, as St. Louis became more industrial, commentators increasingly asked whether the common branches taught in the schools effectively prepared the masses of children for productive labor. This added weight to the familiar complaint that teachers and children were slaves to the textbook, which made the classroom stultifying. It was no wonder, said many critics, that children were often so unhappy at school that they left prematurely, without acquiring the skills necessary to succeed in the workplace.[29]

During Harris's tenure as well, critics had claimed that the schools should better prepare children for work, which was becoming transformed by machines and factories. An exaggerated emphasis on academic subjects, they said, made the schools elitist and uninteresting to the masses of children, and dull teaching and an outmoded curriculum appeared to accelerate the already high withdrawal rates of children. Most pupils left school after age twelve to go to work to help their families survive. As Superintendent Long remarked in 1895 as he stepped down from his post, "Where instruction is lifeless, monotonous, uninteresting, unprofitable, or unreasonably exacting, it becomes the instinctive tendency of a child to withdraw from school as early as he is able to prevail on his parents to let him do so. It is the child's natural protest against faulty treatment."[30] Though Long hoped that schools would become more inviting, he ultimately agreed with Harris that children needed sound academic instruction, not the practical subjects that advocates of manual training, for example, had championed since the 1870s.

Identifying precisely how schools should prepare the young for the "real world" was not easy. Annoyed by those who claimed that the schools taught youth to shun common labor, Harris openly opposed all forms of vocational education, insisting that the common curriculum was the wisest and most practical education for any child. He was suspicious of the grandiose claims reformers had made on behalf of manual training after the Civil War: that it would eliminate industrial alienation, popularize school among the working

---

29. The standard study of the rise of the "new education" and progressivism is Lawrence A. Cremin, *The Transformation of the School: Progressivism in American Education, 1876–1957* (New York: Alfred A. Knopf, 1961).
30. *Forty-first Annual Report of the Board of President and Directors of the St. Louis Public Schools, for the Year Ending June 30, 1895* (St. Louis: Buxton & Skinner Stationery Co., 1897), 47.

classes, and end pupil boredom. Pointing to the hundreds of occupations already listed in the federal census in 1870, Harris doubted that the schools would ever train pupils well for particular jobs, even if they could identify which skills and trades to teach. Schools, he thought, should remain focused on intellectual, civic, and moral training. In addition, Harris worried about the creation of separate school curricula: academic tracks for the favored classes and low-status vocational programs for the poor. Could a ten-year-old, he asked, really know what he wanted to be when he grew up anyway?[31]

The leading advocate of manual training and more practical education was Calvin M. Woodward, who became Harris's lifelong adversary and a worthy rival for influence on the local and national level.[32] On some issues these men were in full agreement. Both opposed training pupils for specific trades and both opposed vocationalism in the public schools. Both also had high praise and respect for the public schools. After that they parted ways. At a lecture at Washington University in 1873, Woodward told the audience that manual training should be part of a wider academic education, since pupils learned in more ways than from books. Invoking the images of Abraham Lincoln and his vice president, Andrew Johnson, he said that in America "there is no limit to the possible social advances of the poor man's child. A nation which bestows its highest honors on a flat-boat man and a rail-splitter of the prairie, and associates with him a man who never went to school, and whose only teacher was his wife, can not expect its sons to fetter themselves by a trade which threatens to tie them down to a life of toil and obscurity."[33] Those trained to be shoemakers, he noted, rarely advanced to higher skilled or better-paying jobs. While Harris dissented, Woodward remained convinced that manual training—defined more broadly than narrow job training—belonged in a modern academic curriculum.

Born in western Massachusetts in 1834, Woodward was a star pupil, winning a scholarship to Harvard and becoming a chaired professor at Washington University at a young age. In the 1880s and 1890s he was one of Harris's chief critics, locally and nationally. The public schools, he claimed, were too oriented around books, causing pupils to drop out prematurely. A professor of mathematics, the sciences, and engineering, Woodward railed against the classical languages and overly humanistic orientation of the city's high school. He was convinced that

---

31. For more on Harris's views on vocational education, see Reese, "Philosopher-King," 165.
32. On the quarrels between Woodward and Harris, see Herbert M. Kliebard, *Schooled to Work: Vocationalism and the American Curriculum, 1876–1946* (New York: Teachers College Press, Columbia University, 1999), 6–12.
33. Calvin M. Woodward, *The Manual Training School* (1879; reprint, New York: Arno Press, [1969?]), 274.

manual training courses would restore dignity to labor, keep young people in school longer, and have the practical benefit of preparing pupils for the real world. He dismissed Harris's prediction that shop class would necessarily become low-status, mostly serving the poor. In addition to opening a preparatory Manual Training School in 1879 at Washington University, Woodward served several terms on the school board, becoming its president in 1900. The school board and school system were increasingly won over to his views, which despite his complaints were ultimately used to justify actual vocational programs.[34]

In his many books, articles, and addresses on the importance of manual training, Woodward argued his case for including practical courses in a modern liberal curriculum. In contrast to Harris's assumption that manual training would become a slippery slope to a narrow vocational education, Woodward believed fairly consistently that liberal and technical education should be unified, not separated, and for everyone, not for different social classes. Like many thinkers of the age, he was convinced that the schools needed to teach eye-and-hand coordination and to train the body as well as the mind. He admired the message of prominent European romantics who argued that children should be active, not passive, in the learning process, and that a diet of words alone did not satisfy the child's hunger to learn.[35]

Thanks to Harris, drawing became a subject in the elementary schools in the 1870s, and the kindergartens, with their emphasis on play and the manipulation of objects, also demonstrated the ways in which the "new education" had slowly entered the schools. Beyond this, neither Harris nor his successor, Edward Long, were willing to venture. The schools faced perennial fiscal pressures, which made Woodward's lobbying on behalf of manual training programs fruitless. Moreover, until the late 1890s, the school board tended to follow the superintendent's lead on this matter, despite some support among its members for manual training. In 1883, Long emphatically denied the charge of some critics that "the tendency of modern education is to create an aversion to manual labor." How could this be true, he asked? After all, "eighty per cent of all the pupils in these schools never advance beyond the mere rudiments of an education. They merely learn to read and write, and to use numbers in problems involving the fundamental processes of arithmetic, and then leave the

---

34. Kliebard, *Schooled to Work*, 112–14. See also Charles M. Dye, "Calvin Woodward, Manual Training, and the Saint Louis Public Schools," *Bulletin of the Missouri Historical Society* 31 (January 1975): 111–35; Peter Sola, "Calvin Milton Woodward," in Altenbaugh, ed., *Historical Dictionary of American Education*, 390.

35. Calvin M. Woodward, untitled essay, *Journal of Education* 22 (December 22, 1885): 411, where Woodward praised two leading European romantic educators, Johann Pestalozzi and Friedrich Froebel.

schools, to pursue the vocation of their parents or some kindred occupation." If former pupils disliked manual labor, he concluded, "[s]ociety is responsible for this, and not the schools. Only two out of every one hundred of our children enter the High School."[36] And so drawing and the various exercises in the city's kindergartens would have to suffice, and public expenditures on manual training, as far as he was concerned, would have to await another day.

The appointment of a new superintendent in 1895 and the election of a reform-minded school board congenial to Woodward's ideas two years later dramatically improved the prospects of manual training. Superintendent F. Louis Soldan praised the value of manual training, and while voicing (like Woodward) his opposition to a utilitarian education, endorsed the "new education" as did the school board, which included Woodward as a member. Woodward had served on the board of education in the late 1870s, and his reelection marked the beginning of a long tenure on the board, a period that witnessed the creation of a number of manual training centers in the elementary schools and more practical courses in the city's high schools. Woodward persisted in his belief that a modernized curriculum would hold students in school longer. He had long espoused this view. As early as the 1870s, for example, Woodward had completed a study on the problem of early school-leavers. Impressed, Harris eagerly published the study, though he disagreed with Woodward's advocacy of manual training as a solution to the problem.[37]

That few pupils stayed in school beyond the ages of twelve or thirteen was a source of concern to most educators after the Civil War. Harris, however, simply argued that children left for many reasons, including economic need, and that weakening the academic curriculum was hardly a sensible solution. By the late 1890s, with Harris a fading memory and Superintendent Long now retired, Woodward and others prepared the pedagogical soil for manual training and practical education. No one ever proved that such programs led to increased enrollments or more pupil satisfaction, stock arguments among their advocates. And, to Woodward's dismay, manual training flowered into narrow vocational programs in the coming decades.

As president of the school board at the turn of the century, Woodward wrote an elaborate report on the virtues of manual training, reiterating themes from his early writings. Once again discounting the fear that manual training would inevitably deteriorate into low-status, nonacademic programs for the poor, Woodward went on to repeat his claim that it represented a solution to the prob-

---

36. *Twenty-eighth Annual Report of the Board of President and Directors of the St. Louis Public Schools, for the Year Ending August 1, 1882* (St. Louis: Slawson & Co., 1883), 106.

37. Dye, "Calvin Woodward," 118.

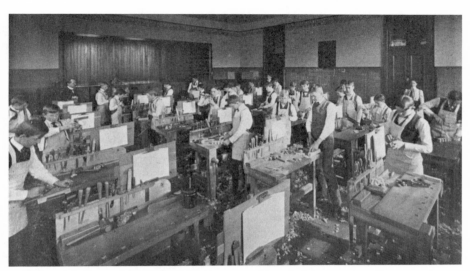

Manual Training Room, Hodgen School (photoengraving from *Forty-fifth Annual Report of the Board of Education of the City of St. Louis, Mo.,* 1899). School board member (and later president) Calvin T. Woodward—whose reputation, like William T. Harris's, extended far beyond St. Louis—championed manual training in the public schools as a logical response to the demands of an increasingly industrial society. *Missouri Historical Society*

lem of high dropout rates. Parents and children, he said, wanted a more "practical education." Shop class for boys and domestic science for girls would make school more appealing and help hold them in school. After sitting through years of book-centered classrooms, children "become tired of the work they have in hand, and they see in the grades above them no sufficiently attractive features to invite them. They become discontented and neglectful; failure follows, they get behind, and then they stop." By the ages of twelve to fifteen, for example, boys often "find the restraints of the schoolroom and grounds very irksome. Many of the things they are required to do seem petty and trivial, and frequent repetitions make them intolerable."[38] Pupils at that age had no desire to memorize more meaningless classroom assignments. Unless manual training and domestic science courses gained greater prominence in a liberal education too long dominated, Woodward believed, by classical languages and other useless subjects, the dropout rates would remain high.

However, the difficulty of realizing Woodward's broad reform ideals had been evident since the establishment of the first manual training program in 1890, when Edward Long was still superintendent. Long, who, like Harris,

38. *Forty-sixth Annual Report,* 27.

tended to give considerable autonomy to local school principals, granted the request of the head of one "colored" school to establish manual training courses, on the condition that he raise private funds. Obadiah M. Wood, principal of the L'Ouverture School, did so and established programs in the use of tools and woodworking by the early 1890s. Other courses followed, for boys and girls, setting an important precedent when Woodward and others gained control over the school board later in the decade.[39]

That manual training programs were first offered to the poorest, most despised social group in the city was an ominous development. Whatever fine distinctions Woodward made between manual training and vocational education were often lost upon other citizens. Earlier in the nineteenth century, manual training was often touted as part of remedial training for juvenile delinquents in youth asylums, children in orphanages, and other unfortunates. African Americans in St. Louis had long fought for access to the public schools; their attendance increased in the 1860s and gained more political support during Reconstruction in the 1870s. Racially segregated schools were nevertheless the norm, and complaints about mistreatment and lack of financial support in these schools remained commonplace in the late nineteenth century.[40] So the initial establishment of manual training in African American schools meant that such courses were always somewhat stigmatized: suitable especially for the poor, and only for other children if kept to a minimum. The experiment in the L'Ouverture School thus spoke volumes about issues of fair treatment in the schools. When manual training courses entered white schools, they became the entering wedge for more vocationalism in the schools. Woodward's dream of making manual training a welcome component of an academic curriculum proved very difficult to achieve.

The public schools of St. Louis evolved in a time of industrial growth, social conflict, and widespread disagreements over the purposes of mass education. Like other residents of their lively city, school leaders quarreled over what knowledge was of the most worth, how children should be taught, and whether a common curriculum best suited the needs of children. By the turn of the century, Harris represented a past that was quickly retreating from view. Few people in the coming years so well defended the need for academic instruction for the masses. Critics who labeled him a conservative and a system builder

---

39. Dye, "Calvin Woodward," 116; Troen, *Public and the Schools*, 157, 166–67.
40. "The Consequences of Racial Prejudice," chap. 4 in Troen, *Public and the Schools*; Gary R. Kremer, "James Milton Turner and the Reconstruction Struggle for Black Education," *Gateway Heritage* 11 (Spring 1991): 67–75; Gersman, "Development of Public Education for Blacks."

would forget his prescient views on the undemocratic nature of vocational education, or his defense of free high schools, coeducation, foreign language training, elementary science instruction, and more flexible promotion policies. With Susan Blow, he had pressed for innovative programs in early-childhood education that proved controversial, yet he persevered over time. Calvin Woodward, struggling for many years in Harris's shadow as a policy maker, helped reshape the schools along a different path, hoping against reality that manual training and practical education would be deemed useful for everyone and not deteriorate into class-based instruction for the city's poorest children.

Change had indeed come to the St. Louis schools. However, despite the enormous growth in the size and complexity of the system, many school practices at the turn of the century still resembled those of an earlier era. The core of the elementary school curriculum was still heavily oriented toward textbooks, study, and recitation. Child-centered pedagogy had made a dent in the system thanks to kindergartens, even though Harris and Blow questioned the assumptions of romantic educators. More important, large class sizes in the elementary grades, which were increasingly taught by women, prevented the widespread adoption of romantic ideals. On all levels of the system, teachers remained the dominant figures in the classroom, and parents often wanted the basics taught in familiar ways. Vocationalism, however, was the new promised land for those convinced that common access to knowledge throughout the system was impractical and undesirable. The advent of ability grouping in the elementary grades and formal tracking in the secondary schools effectively foreclosed the notion of a common school. How children differed, and how schools should train them differently for the capitalist economy, became the mantra of the twentieth century.

By the early twentieth century, the public schools of St. Louis had a secure place in the community. They shared the task of educating the young with a variety of private secular schools and academies, parochial school systems, tutors, parents, churches, and the community more generally. In the shaping of the modern public school system, Harris, Blow, and Woodward placed St. Louis on the educational map; they were remarkable individuals whose influence ranged far beyond the Midwest. The issues they raised were controversial in their time, and remain so today: Do public schools have a civic and moral responsibility to provide all children with a high-quality, academic education? What innovative programs, such as early-childhood education, deserve special support, especially for children living in the most desperate social situations? Finally, as policy makers rush to create more school-to-work programs, is vocationalism a superior path to the future over support for high academic standards and expectations for the city's children?

Part 3

Enriching Life

Theater and Literature

Shaping the Authentic
St. Louis Theater Culture and the Construction of American Social Types, 1815–1860

Louis Gerteis

Henry Shaw's arrival in St. Louis coincided with the onset of an unprecedented economic depression across the United States. Americans soon recovered their exuberant and expansive mood, however, and Shaw himself prospered as a merchant in St. Louis. Nevertheless, the Panic of 1819 held lasting significance for St. Louisans and other Americans. It marked the beginning of what they would come to understand as the "business cycle." The panic signified that the United States had entered a market revolution in which channels of commerce and migration permeated the trans-Appalachian West. As the revolution advanced, Shaw's St. Louis emerged as the gateway to the trans-Mississippi West, a vast terrain within which Americans could invent themselves anew.

Amid expanding markets and territories, Americans suddenly and decisively shed an antitheatrical bias that had prevailed in the land since the colonial era. Theaters thrived as a new physical mobility and an expanding network of economic exchange transformed the United States into a modern, industrializing nation. In a burgeoning American theater, players and managers developed themes and characters that could be made expressive of the American social scene. In theater and in the wider society, St. Louisans and other Americans identified "authentic" social types. The representation and misrepresentation of the theater provided a physical metaphor for uncertainties about authenticity and intention. Inside and outside of the theater, things and people were not always what they seemed to be or were expected to be. In America's market revolution, theater drew much of its energy from a crisis of social representation that arose from the growing anonymity and fluidity of the marketplace. In the theater, where performers and observers joined together in a suspension of disbelief, Americans identified and explored new conventions of social sensibility and sympathy, new ways of presenting themselves and receiving others.[1]

---

1. On the interaction of theater and society, see Jean-Christophe Agnew, *Worlds Apart: The Market and the Theater in Anglo-American Thought, 1550–1750* (Cambridge: Cambridge Univer-

The new popularity of theatrical productions brought audiences to the pit, boxes, and galleries in a manner that mirrored the distinctions that defined social hierarchy in urban America. As the comic actor Joseph Jefferson noted, "theater is divided into three and sometimes four classes." Performers necessarily appealed simultaneously to men and women, whites and blacks, artisans and apprentices, as well as to members of the business and professional classes. Theater managers sought plays and performances capable of filling the inexpensive pit and gallery as well as the more costly dress circle and private boxes. Two St. Louisans, Noah Ludlow and Solomon Smith, became the leading theatrical managers in the West. The partners opened their first theater in St. Louis in the early 1830s in a converted salt warehouse. That structure burned in February 1837 and Ludlow and Smith began construction of the St. Louis Theater, which opened in July of that year. The new theater's interior featured a pit (or "parquette"), a second tier of boxes, and a "gallery" or third tier. As Smith later noted, the gallery provided seating for "colored people." Most of the pit had no seating and provided standing room for white patrons of all classes.[2] Ludlow and Smith remained in the forefront of theatrical entertainment in St. Louis until they dissolved their partnership in the early 1850s.

Performers in theaters like the St. Louis played to the whole house—or, more accurately, played *with* the whole house—because theater in the early nineteenth century held performers and observers in a common state of liminality. As the historian Lawrence Levine has noted, "[T]he gap between the stage and the pit or orchestra, which we have learned to treat as a boundary separating two worlds, was perceived by audiences . . . as an archway inviting participation." Performers understood that an audience defined theatrical success in terms of its own social expectations and anxieties. When performers succeeded, a sense of authenticity could become complete: during a performance of *Othello* in Albany, New York, a boatman in the audience yelled at the actor playing Iago, "You damned-lying scoundrel!" When performers failed, angry audiences drove them from the stage. Still a child when he made his first appearance in St. Louis in the early 1840s, Joseph Jefferson took center stage during a performance on the Fourth of July. The company presented a tableau of Liberty and Independence and planned to have Jefferson sing "The Star-Spangled Banner."

---

sity Press, 1986), particularly 96–100; Richard Sennett, *The Fall of Public Man* (New York: Alfred A. Knopf, 1977), 195–218; and Lawrence W. Levine, "William Shakespeare in America," chap. 1 of *Highbrow/Lowbrow: The Emergence of Cultural Hierarchy in America* (Cambridge: Harvard University Press, 1988).

2. Jefferson quoted in Levine, *Highbrow/Lowbrow*, 25; Solomon Smith, *Theatrical Management in the West and South for Thirty Years* (New York: Harper & Brothers, 1868), 209. See also Louis Gerteis, "St. Louis Theatre in the Age of the Original Jim Crow," *Gateway Heritage* 15:4 (Spring 1995): 34.

He sang the phrase "Oh, say can you see" well enough but then faltered, forgetting the next line. The audience hissed the weeping boy from the stage. As a mature actor, Jefferson warned that "there must be no vagueness in acting," by which he meant that audiences were quick to perceive artifice and that they despised it. "The suggestion should be unmistakable," wrote Jefferson, "it must be hurled at the whole audience, and reach with unerring aim the boys in the gallery and the statesmen in the stalls."[3]

Audiences shaped the substance of theater culture in America, but three individuals in particular laid important foundations for the kinds of theatrical productions that played in early- to mid-nineteenth-century St. Louis: Micah Hawkins (1777–1825), Samuel Woodworth (1784–1842), and James K. Paulding (1778–1860). Hawkins wrote the song "Backside Albany," which, when performed by Andrew Allen in Albany, was said to have introduced blackface song and dance to the stage. Woodworth's patriotic poem "The Hunters of Kentucky," set to music by Noah Ludlow and performed in New Orleans, introduced the prototypical Kentuckian to the stage. A few years later, Woodworth's play *The Forest Rose* introduced the character of Jonathan Ploughboy, the era's most popular stage Yankee. Finally, Paulding's play *The Lion of the West* introduced another popular Kentucky character, Nimrod Wildfire.

The work of Micah Hawkins predated the St. Louis theater but was not irrelevant to it. His "Backside Albany" probably inspired Thomas D. Rice's famous "Jim Crow" material, which was performed in St. Louis in September 1834 and again in August 1848. Also, as the guardian and mentor of the painter William Sidney Mount, Hawkins influenced the development of genre painting in the United States. As I will explain further below, two of the leading practitioners of that art—Charles Deas and George Caleb Bingham—worked in St. Louis and built on Mount's success.

Hawkins stands as progenitor of two interrelated strands of American popular culture. The popularity of blackface theatricality and genre art derived from their ability to portray what audiences received as authentic depictions of everyday life. Theater led the way because it provided the social setting for the construction of authenticity. Genre art followed, providing middle- and upper-class patrons with reassuring depictions of an "authentic" America abstracted from the contentious social setting that spawned them.

Although English comedy had introduced a few blackface characters to the theater as servants, it was Hawkins's "Backside Albany" that offered the first

---

3. Levine, *Highbrow/Lowbrow*, 179–80, 30; Alan S. Downer, ed., *The Autobiography of Joseph Jefferson* (Cambridge, Mass.: Belknap Press, 1964), 41, 283. Jefferson did not return to St. Louis until 1876, by which time he found audiences to be predictably polite. Jefferson quoted in Levine, *Highbrow/Lowbrow*, 25.

example of what became the American minstrel style. In Albany in 1815, Andrew Allen played a Negro character in a play called *Battle of Lake Champlain*. After the play, Allen appeared alone on stage to sing "Backside Albany." Both the play and the song celebrated the American victory over the British at the Battle of Plattsburg, New York, in September 1814. In that fight the American general, Thomas Macdonough, prevailed over the British general, Sir George Prevost. Solomon Smith, who would later join forces with Noah Ludlow in St. Louis, passed through Albany during the 1814–1815 season on his way into the Ohio Valley. There, he saw Allen sing "Backside Albany" and found it original.[4] Allen did not tour as a blackface performer (Smith reported that he later worked as wardrobe manager for the tragedian Edwin Forrest), but "Backside Albany" achieved enough popularity to justify the printing of a sheet music version and the song itself illustrated a new American phenomenon: the use of blackface performers and African American dialect to celebrate patriotic themes:

> Bow wow wow den de cannon gin't roar,
> In Plattburg an all 'bout dat quarter,
> Gub'ner Probose try he han pon de shore,
> While he boat try he luck pon de water,
> But Massa Macdonough,
> Kick he boat in de head,
> Broke he hart, broke he shin, tove he calf in;
> An Gen'ral Maccomb
> Start ole Probose home,
> Tort me soul den I muss laffin.[5]

Blackface performers were not alone in this celebration of patriotic themes. In May 1822, Ludlow—like Smith, not yet arrived in St. Louis—took the stage in New Orleans to sing a song using the words of "The Hunters of Kentucky." Samuel Woodworth had written his poem for the same reason that Micah Hawkins had written "Backside Albany": to celebrate an American victory in the War of 1812. Woodworth's poem celebrated Andrew Jackson's victory at New Orleans. The piece had been published as a broadside in the years immediately after the war and, as the historian John Ward noted, it offered a striking example of the mythmaking process that transformed Jackson into the first American

---

4. Smith, *Theatrical Management*, 138–40.
5. Micah Hawkins, "Backside Albany" (New York: Thomas Birch, 1837), Nineteenth Century Copyright Deposits, Library of Congress, Washington, D.C.

popular hero. The poem celebrated the bravery and skill of the Kentucky riflemen who participated in the defense of New Orleans in January 1815. Their role in the battle had not been a conspicuous one because riflemen were deployed as sharpshooters; the American infantryman fought with a smoothbore musket. But Woodworth's poem told a different story:

> Jackson he was wide awake,
> And was not scar'd at trifles,
> For well he knew what aim we take
> With our Kentucky rifles.
> So he led us down to Cypress Swamp,
> The ground was low and mucky,
> There stood John Bull in martial pomp
> And here was old Kentucky.
> (Chorus)
> O Kentucky, the hunters of Kentucky
> O Kentucky, the hunters of Kentucky[6]

Ludlow did more than put the poem to music. On the night of his performance, he dressed himself in a buckskin hunting-shirt, leggings, moccasins, and a slouched hat and presented himself on stage as a Kentuckian. When the comedy of the evening had ended, Ludlow appeared on stage in his western garb, carrying a rifle. He found the pit of the theater crowded with keelboat and flatboat men who were easily identified by "their linsey-woolsey clothing and blanket coats." He was "saluted with loud applause of hands and feet, and a prolonged whoop, or howl, such as Indians give when they are especially pleased." The stomping and hooting became louder as he began to sing. When he reached the line "And here was old Kentucky," he took off his hat, threw it to the floor, and brought his rifle into position to take aim. "At that instant," he recalled, "came a shout and Indian yell from the inmates of the pit, and a tremendous applause from the other portions of the house, the whole lasting for nearly a minute." He had to sing the song three more times before the audience would let him leave the stage.[7] Like Andrew Allen, Ludlow did not make the stage character he created his stock in trade. But like Allen's early blackface character, Ludlow's Kentuckian soon became a familiar figure on the American stage.

---

6. John William Ward, *Andrew Jackson: Symbol for an Age* (New York: Oxford University Press, 1953), 13–29, 217–18.

7. Noah Ludlow, *Dramatic Life as I Found It* (St. Louis: G. I. Jones & Co., 1880), 237.

Playbill for Dan Marble in *Jonathan in England* and *The Forest Rose* at Ludlow and Smith's St. Louis Theater, April 27, 1849. Samuel Woodworth's highly successful *Forest Rose* presented St. Louis audiences with the influential Yankee caricature of Jonathan Ploughboy. *Missouri Historical Society*

Samuel Woodworth contributed more directly to the American theater when he wrote the musical melodrama *The Forest Rose* (1825). The theater historian Richard Moody has described *The Forest Rose* as America's first theatrical hit, an early-nineteenth-century equivalent of *Oklahoma!* The play met with some success in the late 1820s but became a hit only in the 1830s and 1840s, when several Yankee delineators discovered the comic potential of the country merchant character, Jonathan Ploughboy. Following the lead of James Henry Hackett, who had pioneered the Yankee role, George Handel Hill (heretofore known as a singer of comic and patriotic songs, but soon to be known as "Yankee" Hill) first played Jonathan Ploughboy during the winter season in Philadelphia in 1832. Hackett was in England at the time; when he returned to the United States, he would become best known for his delineation of the Kentucky character Nimrod Wildfire in Paulding's *The Lion of the West*.[8]

St. Louis audiences, themselves familiar through firsthand experience with the varied social types caricatured in such theatrical fabrications, interacted repeatedly with this theatrical fare. A popular Yankee delineator, Dan Marble, played *The Forest Rose* in St. Louis in June 1839, October 1840, October 1841, and every year from 1845 to 1849. As it happened, Marble's last performance in St. Louis, on May 5, 1849, marked the end of his career; he died of cholera on his way to an engagement in Louisville. "Yankee" Hill played *The Forest Rose* in April and May 1848. Hackett played Nimrod Wildfire in *The Kentuckian* in October 1841, May 1845, October 1847, and November 1856. He returned to St. Louis to play a variety of roles in the late 1850s and revived *The Kentuckian* during the Civil War, playing Nimrod Wildfire in November 1861, June 1862, and October and November 1863.[9]

The Forest Rose celebrated the civic virtue of Jeffersonian America, but it did so within a comedy of misrepresentation that drew its humor from the contradiction between the ideal of civic virtue and the reality of a society shaped by competing market interests. The theatrical vehicle for misrepresentation was "Lid" (short for Lydia) Rose, played in blackface to represent a Negro servant. The objects of Rose's misrepresentations were the two antagonists—an English

---

8. Richard Moody, *Dramas from the American Theatre, 1762–1909* (Boston: Houghton Mifflin, 1966), 147–50; [George Handel Hill], *Scenes from the Life of an Actor Compiled from the Journals, Letters, and Memoranda of the Late Yankee Hill* (1850; reprint, New York: Garret & Co., 1953), 52, 74.

9. William G. B. Carson, *The Theatre on the Frontier: The Early Years of the St. Louis Stage* (Chicago: University of Chicago Press, 1932), 286; "Dan Marble—A Last Interview," *St. Louis Reveille*, n.d., Marble Clipping File, Harvard Theatre Collection, Harvard University, Cambridge; Grant M. Herbstruth, "Benedict DeBar and the Grand Opera House in St. Louis, Missouri, from 1855 to 1879" (Ph.D. diss., University of Iowa, 1954), appendix; *St. Louis Republican*, October 14, 1840; October 5, 1841; October 16, 1841; October 26, 1841; May 21, 1844; September 17, 1845.

James Henry Hackett as Nimrod Wildfire (woodcut by T. Sugden, n.d., after A. Andrews). Hackett, one of the best-known theatrical performers of his day, brought his comic character to St. Louis on numerous occasions before and during the Civil War. *By permission of the Folger Shakespeare Library*

aristocrat and the Yankee—whose roles highlighted the problematic nature of civic virtue in an increasingly commercial society.

The play opens with the leading female character, Harriet Miller, longing for the glamour and comfort of city life.[10] Harriet is overheard by the libertine English aristocrat Edward Bellamy and by her suitor, the yeoman farmer William Roseville. Bellamy plans to "dash the native," William, whom he describes as a "bumpkin" and a "mere clodhopper," and take the "sweet simpleton" Harriet with him to the city. Harriet says she will go with Bellamy if her father permits. William is aghast that she would leave him for such dangerous frivolity. Harriet's father, seeing Bellamy for the knave that he is, scolds his daughter: "The girl who would reject the honest heart and hand of an American farmer, for a fopling of any country, is not worthy of affection or confidence."

The comic counterpoint to Harriet's uncertain fidelity is Jonathan Ploughboy's courtship of the deacon's daughter, Sally Forest. Jonathan's efforts to woo Sally are the occasions for a series of jokes played on him. In one instance, Jonathan tells Sally that he is upset because he saw Tom Clover kiss her. "Tom Clover kiss *me*!" exclaims Sally. "A'nt you ashamed of yourself, Jonathan, to tell such a story?" "It is no story," protests Jonathan. "Can you swear to it?" asks Sally. "Yes, on the Bible," replies Jonathan. "Then you would perjure yourself," says Sally, "for it was I that smaked him. Ha, ha ha!" Jonathan protests that Sally is "treating me like a brute," and then delivers what will become his comic tag line throughout the play: "I wouldn't serve a negro so."

Jonathan describes himself as "a little in the marchant way and a piece of a farmer besides." When asked what it is he sells, Jonathan answers, "Every thing." Jonathan is asked if he intends to "shave the natives." "No sir," he answers naively, "every body shaves themselves here." When it is explained to him that "shaving" means driving a "sharp bargain, or what your parson or deacon might denominate cheating," Jonathan replies, "I wouldn't serve a negro so."

At the play's end, Harriet forswears frivolity and independence: those "who sigh for wedded life, / No more your lover's peace molest . . . when you're a wife, I'll bet you my life, / He'll make you repent of the jest." William's sense of order is restored: "nought can annoy a husbandman's joy, / If Heaven but prosper the plough." Bellamy is tricked into kissing the servant Lid Rose, who closes the play with warnings that appearances can deceive.

The Forest Rose's Jonathan Ploughboy provided common ground for all of the Yankee delineators, but the most prominent members of the group sought distinction as well with characters identified directly with themselves. "Yankee" Hill enjoyed his greatest success in this manner as Jedediah Homebred in *The*

---

10. Moody, *Dramas from the American Theatre*, 155–74.

Green Mountain Boy, a play written expressly for him. Hill performed *The Green Mountain Boy,* together with *The Forest Rose,* during his appearance in St. Louis in April and May 1848. An actor billed as "Yankee" Bierce played both pieces in June 1859.[11]

The curtain rises on *The Green Mountain Boy* to find visitors to a country inn encountering Bill Brown, the Negro porter. Jedediah enters "dressed in the usual style of boys about the farms in New England."[12] The action that follows consists of a series of encounters between Jedediah and the other characters, first among them Bill Brown. "Well," says Jedediah, "that's the etarnellest black looking chap I ever see. I never seen one only in the pictur-book; proud as a peacock he was—didn't even look on me." Bill walks by again. "There comes that nigger agin," says Jedediah. "I'll poke fun at him, just to let him see I ain't skeer'd of nobody."

> JED: Halloo, say you, when did you wash your face last; can't tell, can you?
> BILL: Who's you sarsen dere, you know?
> JED: Are you a nigger? I never see a real one, but I guess you be. Ar'nt ye—you?
> BILL: Who's you call nigger?
> JED: Well, I only ask'd you. Why he's mad as a hen a'ready. Did your mother have any more on you?
> BILL: Dere child, you better keep quiet, and mind what you say to me, you little bushwacker; if you am saucy I'll spile your profile, you mind dat now.
> JED: Oh, darn it all, don't git mad, Jack; I only said so out of diviltry, that's all. . . .
> BILL: You mind dat my name am not Jack, I is Bill Brown. I'm a regular rough and tumble nigga fat and saucy, myself, I am, so you better not fool your time wid me, or you get your mother's baby in a scrape.

Jedediah backs off as Bill continues his challenge: "You only just trying to breed a scab on you nose, you up country looking ball face." With that, Jedediah is ready to fight:

> JED: Look here, I'm es good a mind to take right hold and pound your black hide, as ever I had to eat. I'm like the rest of the Yankees. I don't like to begin

---

11. St. Louis Theater Box Book, 1848 Season, Ludlow and Smith Papers, Missouri Historical Society, St. Louis; Herbstruth, "Benedict DeBar," appendix.

12. Passages from *The Green Mountain Boy,* but not the entire play, appear in Hill, *Scenes from the Life of an Actor,* 168–83. *The Green Mountain Boy: A Comedy in Two Acts* (New York: Samuel French & Son, [1860?]) differs from the scenes published earlier in Hill's memoir. The memoir presumably took the scenes from Hill's prompt book. For that reason, my description of the play relies on Hill's memoir.

fightin, but if I once get at it, I don' mind going on with the job no more than nothin. I'm full of grit as an egg is full of meat and yeller stuff, when the dander's raised.

BILL: Well, chicken, you can have a chance.

The stage directions describe Jedediah's behavior: *Brown places himself in a boxing attitude. Jedediah is about to run away, but seeing . . . the hotel-keeper . . . he also stands in an attitude of defence. Jedediah cries out:*

Come on, come on, I'm a thrashing machine, and can be put in motion easy.

The innkeeper orders Bill back to work, and the scene ends.

Neither heroic nor noble, Jedediah Homebred, like Jonathan Ploughboy, presented a rusticity that pitted assurance, vanity, and an eye for the main chance against all of the sophistication and deception of urban society. The blackface characters in these plays provided comic counterpoint for the rustic Yankee. But Bill Brown's character in particular suggests another dimension of the era's theatricality. In St. Louis, as in other river cities, black stevedores and roustabouts interacted daily with white boatmen and laborers. As theater managers, Ludlow and Smith catered to black and white workingmen as well as to the city's tradesmen and businessmen. In the liminal environment of the theater, these representations of racial and regional types contributed to the construction of social identities that could be perceived and received as authentic.

A probable spin-off of *The Forest Rose* was the first broadly popular blackface song, "Coal Black Rose." According to the English comic actor Joe Cowell, who enjoyed a successful career in the United States during the 1820s and 1830s, "Coal Black Rose" was the creation of the New York actor Tom Blakeley, who originated the piece at the Bowery and Park Theaters in the late 1820s. Cowell, and especially his young son, would perform the song. The elder Cowell had never before encountered blackface singing and believed that Blakeley had been the first "to introduce negro singing on the American stage." Cowell was more accurate when he reported that " 'Coal Black Rose' set the fashion for African melodies which Rice for years so successfully followed." Thomas D. Rice probably performed the song during his 1834 appearance in St. Louis when he performed a number of pieces in what one newspaper described as the "popular extravaganza of JIM CROW."[13]

---

13. Joe Cowell, *Thirty Years Passed among the Players in England and America: Interspersed with Anecdotes and Reminiscences of a Variety of Persons, Directly and Indirectly Connected with the Drama during the Theatrical Life of Joe Cowell, Comedian* (New York: Harper & Brothers, 1843), 77–78; Carson, *Theatre on the Frontier*, 140.

The sheet music for the piece indicates that "Coal Black Rose" was sung in two voices, that of Sambo and his unfaithful lover, Rose.

> Lubly Rosa Sambo cum,
> don't you hear de Banjo, tum, tum, tum
> Lubly Rosa Sambo cum,
> don't you hear de Banjo, tum, tum, tum
> (Chorus)
> Oh Rose de coal black Rose
> I wish I may be cortche'd if I don't lub Rose.
> Oh Rose de coal black Rose

Sambo sees his rival Cuffee in the corner, knocks him to the floor, and chases him from the room:

> He jump up for sartin, he cut dirt and run—
> Now Sambo follow arter wid his tum, tum, tum.
> He jump up for sartin, he cut dirt and run—
> Now Sambo follow arter wid his tum, tum, tum.
> (Chorus)
> Oh Rose curse dat Rose
> I wish Massa Hays would ketch dat Rose
> Oh, Rose, you blacka snake Rose![14]

"Coal Black Rose" was the immediate precursor of Thomas D. Rice's "Jim Crow," which combined dance with comic song. Rice himself performed "Coal Black Rose" in Louisville shortly before he appeared for the first time as Jim Crow in 1830.[15] As the era's leading Negro delineator, Rice frequently performed in close proximity to—and at times shared the stage with—Yankee and Kentucky delineators.

James Henry Hackett, the early Yankee delineator who enjoyed his greatest success as a Kentuckian (another regional type that St. Louisans knew not simply as an exotic theatrical character but as a significant element of the local populace), played the leading comic role in James K. Paulding's *The Lion of the West* (1831) and in *The Kentuckian* (1833), a reworking of Paulding's play by

---

14. "Coal Black Rose" (New York: Firth and Hall, n.d.), Sheet Music Collection, Harvard Theatre Collection. "Massa Hays" was a reference to Jacob Hays, the high constable of New York City.

15. See theater ads in the *Louisville Public Advertiser,* April 9 and May 20, 1830.

William Bayle Bernard. *The Kentuckian* permitted Hackett to continue to enjoy popularity as the character Nimrod Wildfire into the 1860s.[16]

The Kentuckian, like *The Lion of the West*, centers on Wildfire's courtship of a visiting English lady, the widowed Mrs. Luminary, who has come to the United States to observe "the domestic manners of Americans." Audiences were reminded of Frances Trollope, whose book *Domestic Manners of the Americans* had been published in 1832. The play opens as a wealthy New York merchant family, Mr. and Mrs. Freeman and their daughter Caroline, prepare for an evening party.[17] Mrs. Freeman harbors aristocratic pretensions and encourages her daughter to favor a Lord Grandby who has been attentive toward her. "In our social system," Mr. Freeman admonishes his wife, "rectitude and talent confer the only title! Why should I not rather give her to a man whose nobility is in his conduct, not his name." In an aside, Mrs. Freeman replies, "What republican infatuation!"

The purpose of the party is to introduce Mrs. Luminary to New York society. In the midst of these preparations Mr. Freeman learns that his nephew, Colonel Nimrod Wildfire, is on his way to New York from Washington. Mrs. Freeman is beside herself, but her husband enjoys Wildfire's native humor as he reads from his nephew's letter: "Let all the fellers in New York know, I'm half horse, half alligator / with a touch of an arthquake, & a bust of a steamboat."

Freeman invites Wildfire to the party to meet Mrs. Luminary. Wildfire construes this invitation as an opportunity to court the widow. Wildfire's rough-hewn manner in wooing Mrs. Luminary becomes the focus of the play and is accented by the comic roles of Caesar, a blackface servant, and "Lord" Grandby. Grandby reveals early on that he is an impostor plotting to marry Caroline to gain control of her fortune. Even as he joins forces with Mrs. Luminary, however, he is no match for Wildfire.

Of Dutch ancestry, Paulding became a prolific author known principally for his humorous histories and depictions of the West. Paulding's writings on race make it clear that he intended audiences to receive Caesar as a comic reproof to Trollope's elitism, and, more broadly, to British antislavery views.[18] Another

16. Moody, *Dramas from the American Theatre*, 149.
17. W. Bayle Bernard, *The Kentuckian. A characteristic drama in 2 acts, (as altered from Hon. J. K. Paulding's original production and first performed at T.[heatre] R.[oyal] Covent Garden in March and in T. R. Haymarket in May 1833)*, Theatre Museum, Victoria and Albert Museum, London, microform. My description of the play follows this manuscript, which bears the inscription "J. H. Hackett Prompt Book."
18. James K. Paulding, *Slavery in the United States* (1836; reprint, New York: Negro Universities Press, 1968); ALH [unidentified acronym], "James K. Paulding," in *Dictionary of American Biography*, ed. Dumas Malone (New York: Charles Scribner's Sons, 1934, 1962), 7:321–22. A Jeffersonian and Jacksonian Democrat, Paulding served in Martin Van Buren's administration as secretary of the navy.

dimension of Paulding's career, his collaboration with Washington Irving in the publication of *Salmagundi* (1807–1808), suggests that he wrote *The Lion of the West* more as an exercise in upper-class satire than as a conscious effort to contribute to a burgeoning popular culture. The reconfiguration of the play as *The Kentuckian* underscores the point that Paulding's character Nimrod Wildfire took on a life of his own in the theater. So did that of Caesar, who played a far more prominent role in *The Kentuckian* than he had in *The Lion of the West*.[19]

As the play continues, Caesar is present when Mrs. Luminary arrives at her New York hotel, and he is quick to correct her impression that all blacks in America are slaves. Only the lowest are slaves, explains Caesar. When he describes himself as a "free black gemmen of color," Mrs. Luminary reveals that her critique of American racial distinctions by no means carries her in the direction of democratic social equality:

MRS. L: Display none of your freedom in my presence!
CAESAR: What marm! you wish to stinguish [extinguish] cibil liberty?
MRS. L: Civil liberty! certainly not, but impertinent liberty, familiarity, sir, so keep your distance.

Later, Caesar brings the widow Colonel Wildfire's "card": a playing card, the king of clubs, with his name on it. Mrs. Luminary asks Caesar if Wildfire is a gentleman. "Don't know marm," he answers, adding that Wildfire described himself as a "horse! Ya! Ya!" Mrs. Luminary, baffled by this expression of western bravado, reasons that Wildfire must have meant that he was "of the Horse," which is to say a "cavalry officer!" Caesar is with Wildfire when he first meets Mrs. Luminary. "Your a Roarer!" says Wildfire to the widow. "Oh, he means *Au*rora, the Goddess of morning!" says Mrs. Luminary in an aside. "What a classical compliment for such a savage!" "In one harness," continues the amorous Wildfire, "we'd make a full team." At this proposal of marriage, Caesar falls to the floor laughing. Wildfire orders him out of the room. "I hab de honor," says Caesar as he runs off laughing. "What," asks Mrs. Luminary after this exchange, "is one free citizen of America averse to one of another color?" "Auge wouldn't stay with him," replies Wildfire. "Indeed! Why not?" asks Mrs. Luminary, setting up Wildfire's curtain line: "Too lazy to shake."

Finally aware of Wildfire's amorous intentions, Mrs. Luminary is desperate to rid herself of him. To this end, she forms a partnership with Grandby. She

---

19. See James N. Tidwell, ed., *The Lion of the West Retitled the Kentuckian, or a Trip to New York* (Stanford: Stanford University Press, 1954).

will help Grandby in his machinations to marry Caroline if he will frighten off Wildfire with the threat of a duel. She also enlists Caesar's help. Caesar will disguise himself as Mrs. Luminary. He agrees to engage in the masquerade "for de ladies, dat am always de dewout wish of dis here palpitating bussom."

Mrs. Luminary's plan falls apart as Wildfire eagerly accepts the challenge to duel. He insists on "going it in the old Mississippi style" with rifles at six paces. "Now, then, back to back," directs Wildfire. At the count of two, the Englishman fires in the air and runs off. Just then, Caesar appears disguised as Mrs. Luminary and falls to his knees at Wildfire's feet, shaking. Wildfire embraces Caesar; "there widow," he says soothingly, "I'll give you a smak." Wildfire lifts the veil and discovers Caesar. "I hab de honor," says Caesar. "Wait till I swallow him whole!" snarls Wildfire. "Yah! Yah Yah!" taunts Caesar as he runs off. Mrs. Luminary, defeated, agrees to be Mrs. Wildfire provided that at the outset of their marriage "you neither call me a screamer nor yourself a horse."

The popularity of the blackface roles in *The Forest Rose, The Green Mountain Boy,* and *The Kentuckian* undoubtedly contributed to the success of the era's leading blackface performer, Thomas D. Rice. After creating the blackface song-and-dance character Jim Crow, he introduced a half-dozen "Ethiopian operas," blackface afterpieces to accompany featured plays. Born in New York City, Rice had first found work as a comic actor with traveling theater companies in the West. His Jim Crow material emerged alongside familiar dramatic and low-comedy fare.[20]

Not surprisingly, Rice developed his Ethiopian operas as elaborations of earlier blackface acts or as burlesques of familiar dramatic fare. "Oh Hush! or the Virginny Cupids" was a farcical skit developed from the song "Coal Black Rose." "Jumbo Jum" was an elaboration of an early Jim Crow song of the same name. Rice's other Ethiopian operas were blackface burlesques, including "Otello," a burlesque of Shakespeare's tragedy; "Virginia Mummy," a parody of the English low comedy *The Mummy;* and "Bone Squash Diavolo," a burlesque of the popular comic opera *Fra Diavolo.* Rice played "Virginia Mummy," "Jumbo Jum," "Otello," and "Bone Squash Diavolo" in St. Louis on several occasions during two weeks in August 1848. A St. Louis stock company revived "Virginia Mummy" for a single performance in January 1861.[21]

"Bone Squash Diavolo" was Rice's most complex theatrical effort. Examined as part of a theater of social representation, it helps to situate early blackface

---

20. See theater ads in the *Louisville Public Advertiser,* June 12, June 14, and September 22, 1830; and in the *Washington,* D.C., *Globe,* October 31, 1833, and January 8, 1838.

21. St. Louis Theater Box Book, 1848 Season, Ludlow and Smith Papers; Herbstruth, "Benedict DeBar," appendix. Rice had planned to perform in St. Louis in April 1838 but sickness prevented his appearance. See Carson, *Theatre on the Frontier,* 243.

"Jim Crow, as Sung by Mr. T. D. Rice at the Theatre Royal, Adelphi" (sheet music cover, London: D'Almaine & Co., n.d.). Rice pioneered a number of African American caricatures, the most influential of which was Jim Crow—the character whose name would attain international recognition as the symbol for an entire set of white cultural assumptions about African Americans and their standing in American society. *The Harvard Theatre Collection, The Houghton Library*

entertainment as one element of a broader effort to represent distinctive American types. The original comic opera, written by Daniel Francois Esprit Auber in 1830, featured a hero, Lorenzo, who captures the bandit leader Fra Diavolo and in the process wins the hand of the woman he loves, Zerlina, whose father had planned her marriage, against her romantic desires, to a wealthy farmer. In Rice's Ethiopian burlesque, Bone Squash sells his soul to the devil to win the hand of Junietta Ducklegs. But Junietta wants to wed a dandy, Spruce Pink, rather than the chimney sweep Bone Squash. Bone Squash's marriage scheme collapses as he struggles to keep out of the clutches of the devil. In the burlesque, Rice transported Auber's opera to the streets of New York and, through racial masquerade, engaged the mysteries of markets and the markings of class in an increasingly commercial society.[22]

The curtain rises on "Bone Squash" to reveal a grogshop at dawn. Caesar (a porter) awakens in a wheelbarrow as Jim Brown (a fiddle player), Mose (a bootblack), and Juba (a whitewasher) join him. Brown offers expansive malapropisms as he claims for himself as a fiddler a status superior to the others: "I cannot help my facetious humor to meditate how Natur hab rabished her comeasticle endowments upon de human family." They all dance a shuffle as Brown sings: "I am de child of genus, / And my name's Jim Brown, / I fiddle at de Five Points, / And all about de town." Bone Squash enters with the tools of the chimney sweep, a coil of wire and a brush, over his shoulder. He complains that he has been all around town and found no work. "I wish de debbil had de man what first discovered de coal fires," says Bone Squash, who insists that coal-burning has diminished his livelihood:

> I doesn't understand de chemistry ob de 'gredients 'nuff to disqualify dem, but no sooner den de coal smoke get in de chimney, den he right away emigrate out ob de top, and 'waporates into de native element like—He doesn't adherify to de sides like de wood smoke does.

"Oh, I wish I could sell myself to de debbil," sings Bone Squash. Suddenly the devil appears from a hogshead in a flash of fire: "How de do? Rather guess I heard you say you wanted to see me. Well, here I am, piping hot!" Bone Squash falls on his back, shaking violently: "Is you de debbil?" he asks. "Yes," replies the devil, "a real genuine Yankee devil, or, a devil of a Yankee; you may have me either way by paying the discount." The devil tells Bone Squash that some of the best people of New York could not get by "without they have a whole devil

---

22. *Fra Diavolo: Comic Opera in Three Acts* (New York: Academy of Music, n.d.); *Bone Squash. A Comic Opera* (New York: Frederick A. Brady, n.d.).

to themselves." "Well, how many does de nigger hab?" asks Bone Squash. That all depends, replies the devil: "If he belongs to the temperance society, we put a devil to every two . . . but if he's a real double distilled swell head, he'll burn a fortnight without any fuel, and can take care of himself."

Bone Squash agrees to sell himself, but he expects a good price:

> Look a here, Mr. Yankee debbil; I'm no common nigger what you meets wid round de markets and de wharves. I am a gemmen ob color, what libs by de sweat ob de chin, as de poet says; and if you buy me, you must just crowd steam and come up to de landing, pretty saucy. You see, honey, I'm a free nigger.

The devil offers two hundred dollars. Bone Squash says he is worth four hundred. They settle on three hundred as the devil tells the audience that "he's worth five hundred, if he's worth a bushel of peas." Bone Squash gives the devil his bond. The devil gives Bone Squash a check (he is well known to all of the brokers, he assures Bone Squash). The two seal their deal with a drink and sing a duet.

In the next scene, an extravagantly dressed couple—Spruce Pink and Junietta Ducklegs—walk down Broadway commenting on the heat. Like Jim Brown, their language is pretentious and filled with malapropisms. Junietta explains to her escort why it is so hot:

> My lub, de reason am berry perspicous. Yesterday, de inky clouds overhung de earth as black as de smoke around de bake-house; darfore, de heat could not perforate; but, in de rotundity ob natur, dem inky clouds hab all distinguished away, and de heat dat we ought to hab come down yesterday, we 'spect we had to-day, consequently, we hab two days' heat in one.

"I did not tink ob dat," responds Spruce, suggesting that they get out of the sun before they are "tanned as black as de common white folks."

The scene shifts to Junietta's parlor, where she is shocked to learn that fortune has smiled on Bone Squash and that her father intends to marry her to the chimney sweep. In despair, she delivers a burlesque lament:

> It nebber can—it nebber shall be. Sooner den be Mistress Squash, dat hateful, vulgar name, I exile to Siberia, and live among de Indians. I make de solemn vow to lub Spruce Pink, dey shall not jerk me from him. *(Faints)*

Jim Brown arrives with his fiddle and the marriage ceremony begins. Suddenly the devil appears from a trapdoor. "What do you say to emigrating?" he asks

Bone Squash. "I can't go jist yet," replies Bone Squash, "I am gwan to be married." "So much the better," says the devil, "bring your wife along."

In the final scene, everyone gathers around a balloon being prepared for ascent by two devils with bellows. Bone Squash enters, singing, "Save me! save me! save me!" The rest sing in reply: "Oh, de nigger must be crazy, / It's berry plain to see: / De Debbil's comin' after him, / He can't get free." The devil puts Bone Squash in the basket of the balloon, and the ascent begins amid an explosion of fireworks. Bone Squash cuts one of the ropes holding the basket to the balloon; it tilts and the devil falls out. Bone Squash continues his ascent, according to the stage directions, "throwing out his shoes, hat, etc." As flames shoot up, the devil falls through a trapdoor and the curtain falls on the final tableau.

Two themes can be fruitfully explored here, both of which would have resonated with audiences in a city growing as quickly as St. Louis. The first is the commodification of labor. The devil in Rice's burlesque is a Yankee, a stockjobber in the language of the day, who deals in paper and is well known to all the brokers. His financial dealings with Bone Squash are representations of a placeless market and a process of continuous exchange that is at once the source of optimism and uncertainty in American society. The devil buys Bone Squash for what he says is a fraction of his true value. But there is no standard of value (in labor, in commodities, or in Bone Squash's soul) except that which the devil enunciates. The devil would also have Bone Squash "emigrate" and hopes to take Junietta in the bargain, an ambiguous reference to the efforts of the American Colonization Society to settle free blacks from the United States in Liberia. Significantly, the Yankee devil does not have his way. Bone Squash does not get Junietta, but neither does he emigrate with the devil.

A second theme relates to class. Bone Squash's trade—like those of his companions—is a lowly but honorable one and by his own lights it makes him a gentleman. Bone Squash is a sympathetic as well as a comic character in his successful flight from the devil. The objects of ridicule in the farce are Spruce Pink and Junietta. Junietta's preference for Spruce Pink over Bone Squash is an inversion of the heroine's familiar capacity to distinguish fundamental honesty and courage from sham nobility and to seek romantic love over wealth and station. Junietta is spared marriage to the "vulgar" Bone Squash, but she can hardly be considered a worthy prize for him. In the end, Bone Squash is triumphant, making good his escape from the all-consuming Yankee devil and the commodification of self that the devil represents.

Just as audiences participated in shaping these theatrical types, they carried them into the street in public masquerade. In February 1847, St. Louisans gathered for a grand parade to celebrate the founding of the city. All of the city's fire companies, schools, societies, and orders displayed themselves in characteristic attire along the flag- and flower-decorated route. A twenty-foot replica of the *General Pike* (the first steamboat to arrive in St. Louis, in July 1817) occupied a prominent position near the head of the procession. An assembly of boatmen and boys followed. Members of the Hunting Club marched in hunters' costumes. Members of the Hibernian Charitable Society identified themselves with green sashes. They were followed by a large group in masks riding in carriages and on horseback. "The grotesqueness of the dresses and the variety of the characters," noted a commentator, elicited the excitement and amusement of the crowds that lined the route. The masquerade represented "male and female, white and black, old and young, Yankee, French and other characters." Here, too, St. Louis was representative of urban America. In a masquerade in Philadelphia in 1834, for example, a band of "Indians, hunters, Falstaffs, Jim Crows and nondescripts" marched behind a mock militia. The social types that had become familiar fare in the theater became part of the popular perception of an American self.[23]

As mentioned, theatrical delineators appealed simultaneously to the boys in the gallery and the statesmen in the stalls. In upper- and middle-class parlors, by contrast, class-specific representations of authentic social types took shape. The proliferation of published sheet music in the 1840s and 1850s provided musical arrangements for the pianoforte of songs first made popular in the theater. Frequently adorning the sheet music were lithographic covers that drew their inspiration from the theater. These covers presumably helped to shape musical interpretations in the home, just as the original performances had within a public setting. The painters and draftsmen who have since become known as "genre artists" similarly applied theatrical inspiration to the private articulation of upper- and middle-class social expectations and anxieties. Like the broader theater audiences, their well-off patrons relished what they received as authentic renderings of social types. As the art historian Elizabeth Johns has noted, the authenticity of images produced by such artists as William Sidney Mount, Charles Deas, and George Caleb Bingham encompassed the viewer's own "ideological underpinnings"; a successful painter therefore "could be said to be an entrepreneur of the viewer's ideologies."[24]

---

23. *Report of the Celebration of the Anniversary of the Founding of St. Louis, on the Fifteenth Day of February, A.D. 1847* ([St. Louis]: Chambers & Knapp, 1847). For masquerade and theater in Philadelphia, see Susan G. Davis, "'Making Night Hideous': Christmas Revelry and Public Order in Nineteenth-Century Philadelphia," *American Quarterly* 34 (Summer 1982): 188; Susan G. Davis, *Parades and Power: Street Theatre in Nineteenth-Century Philadelphia* (Philadelphia: Temple University Press, 1986), 77.

24. Elizabeth Johns, *American Genre Painting: The Politics of Everyday Life* (New Haven: Yale University Press, 1991), xii.

The relationship between the perceived authenticity of such images and the theatrical conceits of the day was by no means coincidental. One of the period's most accomplished genre artists, William Sidney Mount, had been introduced to comic art by his uncle, Micah Hawkins, with whom he lived as a boy following the death of his father. Mount became popular among wealthy patrons for his paintings of Yankee farmers. While Hawkins's work had been theatrical and public, Mount's work was privately displayed, but the influence of the theater is unmistakable. *Bargaining for a Horse* (1835), depicting the meeting of a young man and an older farmer, was commissioned by Luman Reed, a wealthy New York merchant from an impoverished rural background. Reed was a friend and admirer of the comic actor James Henry Hackett, whose character Jonathan Ploughboy set the style for the Yankee delineators of the early nineteenth century and probably informed Mount's painting as well.[25]

African American characters were of equal importance to Mount's work. In his first genre painting, *Rustic Dance after a Sleigh Ride* (1830), Mount included three black servants on the margins of the painting, one of them playing the fiddle for the white dancers, another tending the fire, and the third, a coachman, looking on from the doorway. Mount would soon make African American figures more central to his work, always in the manner of a theatrical *mise-en-scène*. In one of his most famous paintings, *Farmers Nooning* (1836), Mount placed an African American figure, sensual, supine, and napping, on a mound of freshly cut hay, while a mischievous boy in a tam-o'-shanter prepares to tickle him in the ear with a straw. "Ear tickling," Johns notes, meant "filling a vain listener's mind with promises." The tam-o'-shanter was a familiar visual emblem of British abolitionism. The boy, therefore, represented abolitionism. His actions are about to arouse the peaceful black man and destroy the tranquility of the moment. Yet something in the theatrical quality of the scene helps to defuse its volatility. Unlike the virulent antiabolitionist cartoons of the era, Mount's painting does not induce its viewers to fear abolitionists or the objects of their philanthropy.[26]

Mount's *The Power of Music* (1847) once again presented blacks and whites in theatrical juxtaposition; while two older white men in a barn watch a young man playing the fiddle, a third man stands unseen by them outside of the barn door. He is an appreciative black listener, visibly older than the performer. Did he teach the white man how to play? Is he amused, hearing in the white man's music the melody and phrasing of a African American song? The answer is not obvious. In any case, the musical performance itself, like the suggestion of ongoing action among clearly identifiable social types, relates Mount's

25. Ibid., 24, plate 3 [n.p.].
26. Ibid., 33–35, plates 2, 4 [n.p.].

work back to the theatrical performances with which his viewers were already familiar.

By the 1840s, however, the popularity of Mount's rustic images had begun to fade. Amid agitation for the annexation of Texas, the excitement of the Mexican War, and the subsequent expansion of the United States to the Pacific Ocean, popular attention turned westward. Two St. Louis artists, Charles Deas and George Caleb Bingham, gained national prominence for their depiction of western characters and scenes. Both continued to draw on theatrical convention in their work.

Deas's rendering of *Long Jakes* (1854), for example, retained aspects of the familiar, rough-hewn Kentuckian at the same time that it presented viewers with a romantic view of the trans-Mississippi West now associated with men such as Kit Carson, recently made famous as John C. Fremont's guide in his well-publicized western explorations. The first print of *Long Jakes,* published in 1846, remained true to Deas's theatrical depiction of the uncouth Kentuckian, at once self-reliant and uncivilized. But romanticism triumphed, as a much more widely circulated 1855 print rendered *Long Jakes* unequivocally handsome and heroic.[27]

Deas's fellow St. Louisan George Caleb Bingham also drew directly from the theater. *The Jolly Flatboatmen* (1846) depicts a boatman dancing a jig as another plays the fiddle and a third uses a tin plate as a tambourine. Viewers would likely have been reminded of such blackface songs as "The Boatmen Dance" and "Gumbo Chaff"—the difference being, of course, that Bingham's boatmen were not in blackface. Indeed, Bingham did not make significant use of African Americans in his genre paintings. In *The Verdict of the People* (1853–1854), one of his political scenes, a black man appears in the left foreground pushing a wheelbarrow to collect a drunk who has fallen in the street. But, in contrast to Mount and to Richard Caton Woodville, whose *War News from Mexico* (1848) also treated political themes, Bingham's black man is peripheral to his painting. In *Jolly Flatboatmen in Port* (1857), Bingham made more significant use of an African American figure. Again Bingham painted a configuration of white male figures: a jig dancer, a fiddle player, and a tin-pan-tambourine player. This time, however, he surrounded them with men on a landing engaged in various activities. Noticeably disengaged from labor is the black man, dressed in tattered clothing, smiling and looking at the dancer. In contrast to the radiant smiles on the faces of the dancer and tambourine player, however, the black man's grin is slack and his hands hang awkwardly and without purpose in front of him.

27. Ibid., 66–67.

Long Jakes; the Rocky Mountain Man, 1844, Charles Deas. In Deas's image, the familiar characteristics of the theatrical Kentuckian worked their way into a newer idealization of the self-reliant western scout. *Denver Art Museum Collection: Funds from Collector's Choice 1999 and T. Edward and Tullah Hanley Collection by exchange, 1998.241. © Denver Art Museum 2002*

Jolly Flatboatmen in Port, 1857 (oil on canvas by George Caleb Bingham). Bingham's paintings nostalgically recalled life along the preindustrial river—and portrayed African Americans as observers of, rather than participants in, the sorts of roles that performers might once have staged in blackface. *The Saint Louis Art Museum*

Unlike the African American figure in Mount's *The Power of Music*, Bingham's is a fawning onlooker, overawed by the performance.[28]

Although Bingham distanced his white performers from blackface entertainment, his work paralleled the development of theatrical fare. His boatmen and the landscapes in which he placed them were nostalgic renderings. Steamboats had long since replaced flatboats and keelboats on western waters. Similarly, Bingham's genre paintings depicting political scenes, notably *The County Election* (1852), offered humorous but nostalgic views of urban spaces free of crowded omnibuses, steam-powered factories, and other signs of the industrialization that had begun to transform St. Louis and other northern cities. Bingham's election scenes also depicted a homogeneous electorate free of the ethnic divisions that put native Americans and immigrants in hostile political camps. Bingham painted his scenes of democratic indulgence at a time when St. Louis experienced two riots involving German and Irish immigrants and nativists.

28. Ibid., plate 9 [n.p.]; figure 37, 139; plate 1 [n.p.].

In the theater, blackface performances became the dominant mode for expressions of nostalgia. Although the popularity of genre painting declined in the 1850s, Eastman Johnson's *Negro Life in the South* (1859) received a warm reception precisely because it embraced the nostalgia of the new minstrel stage. A New York critic imagined that the black banjo player in the painting was entertaining his listeners with the sentimental minstrel song "Poor Lucy Neal." The painting would also bear the title *Old Kentucky Home,* a direct reference to Stephen Foster's 1853 minstrel song.[29]

The 1850s marked the end of a theatrical era. The class divisions and antagonisms that were so brutally bared in the Astor Place riot in New York City in 1849 signaled that elites would no longer submit to what they regarded as the cultural tyranny of the masses. At the same time, the staging of Harriet Beecher Stowe's *Uncle Tom's Cabin* heralded the dawn of a new age of mass entertainment characterized by large theatrical "combinations" (companies that no longer relied on local stock players but traveled from city to city with their own players, scenery, and costumes) and extravagant display. When Solomon Smith wrote his reminiscences in the late 1860s, he lamented the decline of traditional drama. The people, he wrote, had had "Shakespeare fumigated out of them by red fires and blue blazes." Another critic noted that the theater of the late nineteenth century provided mass entertainment "without any contributing effort on one's part." The new theater of extravaganza benefited from the "spectator's more passive attitude."[30]

George L. Aiken's dramatization of *Uncle Tom's Cabin* gave rise to the first theatrical combination, a company headed by George C. Howard, whose wife was Aiken's cousin and whose daughter, Cordelia, became famous in the role of Little Eva. Stowe herself had inherited a strong antitheatrical bias from her father. In 1843, Lyman Beecher had denounced the theater as "the great social exchange where sinners of all grades, colors and description assemble to barter away and sell their immortal souls." In St. Louis, the Reverend William Greenleaf Eliot echoed these sentiments. In an 1848 lecture on entertainment, Eliot warned that the theater was "surrounded by incidental evils." Although the Unitarian minister saw beauty in Shakespeare and conceded that "the theatre mingles with . . . entertainment some elements of instruction and intellectual enjoyment," he regarded theater in general as an idle amusement. Like the novel, it should be avoided. Young men, cautioned Eliot, should spend their

---

29. Ibid., 130, plate 15 [n.p.].
30. Smith, *Theatrical Management,* 238; John Russell David, "The Genesis of the Variety Theatre: 'The Black Crook' Comes to St. Louis," *Missouri Historical Review* 64 (January 1970): 145.

leisure time reading history and biography. When Asa Hutchinson of the abolitionist Hutchinson Family Singers sought Stowe's permission to adapt her novel to the stage, Stowe refused, saying that "the world is not good enough yet for it to succeed."[31] Nevertheless, Stowe's novel displayed a theatricality of its own that made Aiken's unauthorized staging effective and ultimately more popular than the novel itself. Stowe's Uncle Tom embodied the sentimentality that had come to characterize blackface entertainment in the late 1840s. The Kentucky slave, Sam, spoke with minstrelsy's familiar dialect and malapropisms. St. Clare introduced Topsy to Miss Ophelia as "a funny specimen in the Jim Crow line" and Uncle Tom's Kentucky owner referred to Eliza's son in the same manner. Finally, Simon Legree can be viewed as the Yankee devil besotted by "shaving" to the point of moral debasement and madness.[32]

In 1855, George Howard's company took its *Uncle Tom's Cabin* production on the road. In 1858 they arrived in St. Louis and opened at Ben DeBar's New St. Louis Theater on May 24.[33] The *Missouri Democrat* recorded the performance with enthusiasm but the proslavery *Republican* found in it "more psalm-singing, moral lectures and general parson-like primness than we have ever seen in a stage piece before." In her memoir, Cordelia Howard recalled a reserved reception for her father's company. There were no "disagreeable incidents," she wrote, but "we never gave 'Uncle Tom' any further South."[34]

Earlier versions of *Uncle Tom's Cabin* had played in St. Louis to a more positive response. The People's Theater offered *Uncle Tom's Cabin As It Is* in Jan-

---

31. Beecher quoted in Smith, *Theatrical Management*, 175; William Greenleaf Eliot, *Lectures to Young Men, Delivered in the Church of the Messiah* (St. Louis: Republican Printing Office, 1852), 41; Stowe quoted in Eric Lott, *Love and Theft: Blackface Minstrelsy and the American Working Class* (New York: Oxford University Press, 1993), 273 n. 5.

32. Harriet Beecher Stowe, *Uncle Tom's Cabin, or, Life Among the Lowly* (1852; reprint, New York: Bantam Books, 1981), 3, 47, 74, 237, 342–43.

33. When Ludlow and Smith dissolved their partnership and retired in 1851, the original St. Louis Theater disappeared to make room for a federal customhouse and post office. Thereafter, theater manager John Bates operated the Bates Theater, also known as the New St. Louis Theater, until he sold it to Ben DeBar in 1856. A native of London, DeBar began performing in the United States in the mid-1830s and became popular in the role of Falstaff and other boisterous comic characters. He operated the New St. Louis Theater, also known as DeBar's Opera House, until he retired in 1873. See J. Thomas Scharf, *History of St. Louis City and County, From the Earliest Periods to the Present Day: Including Biographical Sketches of Representative Men* (Philadelphia: Louis H. Everts & Co., 1883), 1:980–81. I am indebted to Lori Singer, "St. Louis Theater: From Lincoln's Election through the Battle of Wilson's Creek," a 1997 University of Missouri–St. Louis graduate seminar paper.

34. *St. Louis Missouri Republican*, May 27, 1858; Cordelia Howard MacDonald, "Memoirs of the Original Little Eva," *Educational Theatrical Journal* 8 (1956): 277. In an engagement that began May 24 and ended June 5, 1858, the Howard company played *Uncle Tom's Cabin* on four consecutive nights (May 24–27). Perhaps sensing that the city did not fully sympathize with the play's antislavery theme, they billed Cordelia Howard as the star and called the play *The Death of Eva*. See Herbstruth, "Benedict DeBar," appendix.

uary 1854 and received favorable reviews in the antislavery *Democrat*.³⁵ The St. Louis actor Joseph M. Field produced an antiabolitionist version, *Uncle Tom's Cabin; Or Life in the South As It Is*, at the Variety Theater on April 29 and May 2, 1854. The playbill described the piece as having been written by "Mrs. Harriet Screecher Blow," a reference to Dred Scott's benefactor in St. Louis, Henry Taylor Blow. Field's production played to full houses on both nights.

Aiken's play enjoyed greater popularity in St. Louis well into the post–Civil War era. The Park Theatre Combination staged seven performances in March 1878. The Palmer and Company combination staged sixteen performances in May of the same year. By then, the antebellum delineators and their audiences had long since passed from the scene.³⁶

In the public sphere of theater and in the more private world of genre art, St. Louis contributed significantly to the development of popular culture in antebellum America. The city had been particularly prominent in the decade of the 1840s, when Thomas D. Rice, "Yankee" Hill, Dan Marble, and James Henry Hackett were popular and familiar theatrical performers in the city. The extraordinary popularity of *The Forest Rose* and *The Kentuckian* suggests that St. Louis audiences received and responded to Marble's Jonathan Ploughboy and Hackett's Nimrod Wildfire with particular enthusiasm. Poised precariously over a widening sectional divide, St. Louis proved less hospitable to the theatrical extravaganza of *Uncle Tom*. At the same time, however, the city's genre artists, Deas and Bingham, found that the tense climate of North-South sectionalism—as acutely felt here as it was in any city in the United States—contributed to a growing demand for idealized images of westerners, the newest authentic Americans.

---

35. *St. Louis Democrat*, January 11 and 13, 1854. For information on these productions, I am indebted to Annette Thompson, "Uncle Tom's Cabin as a Social Document: The Construction of Race in St. Louis Theatre," a 2000 University of Missouri–St. Louis independent study paper.

36. Herbstruth, "Benedict DeBar," appendix.

Themes and Schemes
The Literary Life of Nineteenth-Century St. Louis

Lee Ann Sandweiss

Cities, like people, have character. As with people, a part of that character is innate or inherited, while another part develops only through interaction with others. In the imprint left upon it by the explorers, settlers, longtime inhabitants, and tourists who crowd its streets, a city slowly gains the complex personality that it lacks as a mere site upon the land. Through the thoughts and words of those who have experienced the place, and who have cared enough to record their impressions on paper, that personality begins to cohere around common images: civilized or wild, inviting or frightening, thriving or dying, progressive or backward. St. Louis has always possessed an unusually rich and ambiguous amalgam of such images. From the first, almost incidental, accounts of its site through the highly self-conscious writing of the late 1800s, the city developed as a place of competing literary images. Throughout the nineteenth century, that conflicted character would spur a common thread of concern among the city's literary observers for St. Louis's civic welfare.

The first literary mention of the region around St. Louis is less notable for what it says than for what it overlooks. In 1673, nearly a century before Pierre Laclede's party touched the west bank of the Mississippi, the French Jesuit Jacques Marquette accompanied his countryman Louis Jolliet on a voyage to verify the existence of the "Mesippi" and to bring Christianity to the Indians of the southern regions of New France. Marquette's journals of the 1670s, filled with wonder and wariness, set an otherworldly precedent for subsequent literary descriptions of the region that now encompasses St. Louis: "From time to time, we came upon monstrous fish, one of which struck our canoe with such violence that I thought that it was a great tree, about to break the canoe to pieces. On another occasion, we saw on the water a monster with the head of a tiger, a sharp nose like that of a wildcat, with whiskers and straight erect ears; the head was gray and the neck quite black."[1]

Indeed, Marquette's perception of the so-called Piasa Bird, painted high on the bluffs on the east side of the river near what is now Alton, Illinois, suggested his fear of having traveled not just to unexplored territory but to another world:

> While skirting some rocks, which by their height and length inspired awe, we saw upon one of them two painted monsters that at first made us afraid, and upon which the boldest savages dare not long rest their eyes. They are as large as a calf; they have horns on their heads like those of deer, a horrible look, red eyes, a beard like a tiger's, a face somewhat like a man's, a body covered with scales, and so long a tail that it winds all around the body, passing above the head and going back between the legs, ending in a fish's tail. Green, red, and black are the three colors composing the pictures. Moreover, these two monsters are so well painted that we cannot believe that any savage is their author; for good painters in France would find it difficult to paint so well,— and, besides, they are so high up on the rock that it is difficult to reach that place conveniently to paint them.[2]

Nowhere, in this roster of wonders, did the site that became St. Louis gain so much as a casual notice.

Through the eighteenth century, others would follow Marquette's lead in attempting to describe the essence of the landscape and culture of what is now greater St. Louis. Most of the "literature" of this era consisted of travel narratives and journal entries by French explorers such as Charlevoix, Hennepin, La Salle, and later, Nicolas De Finiels, who also produced one of the most detailed early maps of the St. Louis settlement. The most thorough account of eighteenth-century St. Louis is actually a piece of nineteenth-century literature: the fragmented "Narrative" of city cofounder Auguste Chouteau, first discovered in 1855, which presents a detailed, if slightly self-aggrandizing and romanticized, version of St. Louis's founding in the winter of 1763–1764. Although the veracity of the document and date of its composition are less than certain, Chouteau's narrative stands as the most frequently cited firsthand account of that historic event:

---

1. Jacques Marquette, "Le premier voyage qu'a fait le P. Marquette vers le nouveau Mexique.— Journal incomplet, adressé au R. P. Dablon," in Reuben Gold Thwaites, ed., *Voyages* (Ann Arbor, Mich.: University Microfilms, 1966).
2. Ibid.

"Founding of St. Louis, 1764" (chromolithograph by National Colortype Co. after a painting by E. Cameron in the *St. Louis Globe-Democrat,* February 28, 1902). Embellished through literary accounts in the nineteenth century and artistic images in the twentieth (such as this one, commissioned by the Anheuser-Busch Co.), the establishment of St. Louis by Pierre Laclede and his teenaged assistant, Auguste Chouteau, proved to be the city's most enduring legend. *Missouri Historical Society*

[Laclede] was delighted to see the situation (where St. Louis at present stands); he did not hesitate a moment to form there the establishment that he proposed. Besides the beauty of the site, he found there all the advantages that one could desire to found a settlement which might become very considerable hereafter. After having examined all thoroughly, he fixed upon the place where he wished to form his settlement, marked with his own hand some trees, and said: "Chouteau, you will come here as soon as navigation opens, and will cause this place to be cleared, in order to form our settlement after the plan that I shall give you."

I arrived at the place designated on the 14th of February, and, on the morning of the next day, I put the men to work. They commenced the shed, which was built in a short time, and the little cabins for the men were built in the vicinity. In the early part of April, Laclede arrived among us. He occupied himself with his settlement, fixed the place where he wished to build his house, laid a plan of the village which he wished to found (and he named it St. Louis, in honor of Louis XV, whose subject he expected to remain, for a long time;—he never imagined he was a subject of the King of Spain).[3]

Chouteau's "Narrative," while recalling the community at the moment of its creation, appeared at a time when a growing number of writers were projecting their literary imaginations across the tabula rasa of St. Louis's youthful facade. Many more would do so before century's end. Yet the spectrum of literature on nineteenth-century St. Louis, taken broadly to include all forms of writing, does reflect a measure of consistency. It is framed by two important, and complementary, documents from opposite ends of the century, each a summons—a challenge to the citizenry—to assume responsibility for helping St. Louis realize its potential as a cultural as well as commercial center. The first, intended especially for the citizenry of a newly Americanized Upper Louisiana, was written in March 1804 by Captain Amos Stoddard, the new commandant of the territory. The second, William Reedy's 1899 essay "What's the Matter with St. Louis?" addressed a populace considerably more accustomed to city life but no less in perceived need of direction in the responsibilities of citizenship.

In his "Address to the People of Upper Louisiana," Amos Stoddard not only informed the citizens of their good fortune at now joining the United States, he also reassured them that their customs, religions, and basic ways of life would not be tampered with. At the same time, he reminded them that citizenship carried with it "many reciprocal duties . . . and the prompt and regular performance of them is necessary to the safety and well fare of the whole. No one can plead exemption from them. . . . [E]very soldier is a citizen, and every citizen a soldier." Stoddard boldly predicted that, if the citizens honored the law of the land and worked to maximize the benefits of the area's natural bounty, "Upper Louisiana will, in all probability, soon become a star of no inconsiderable magnitude in the American constellation."[4]

As rationally as Stoddard framed the essential character of the new American territory, it would be many years before writers ceased to capitalize on

---

3. Auguste Chouteau, "Narrative of the Settlement of St. Louis" *Missouri Historical Society Collections* 3:4 (1911): 47–59. The original French-language manuscript is in the holdings of the St. Louis Mercantile Library. For more on the context of the discovery and publication of the "Narrative," see Eric Sandweiss, *St. Louis: The Evolution of an American Urban Landscape* (Philadelphia: Temple University Press, 2001), 241 n. 8, 248 n. 5.

4. Amos Stoddard, "Address to the People of Upper Louisiana, 10 March 1804," *Glimpses of the Past* 2:6–10 (May–September 1935): 88, 89, 91.

(and, at times, to exaggerate) the city's exotic frontier image. In 1817, for example, there appeared in Boston a book modestly entitled *The Narrative of the Captivity and Sufferings of Mrs. Hannah Lewis, and Her Three Children, Who Were Taken Prisoners by the Indians, near St. Louis on the 25th of May, 1815, and among Whom They Experienced All the Cruel Treatment Which Savage Brutality Could Inflict—Mrs. Lewis, and Her Eldest Son, Fortunately Made Their Escape on the 3rd of April, Last, Leaving Her Youngest Two Children in the Cruel Hands of the Barbarians.* A typical example of the captivity-narrative genre that flourished during the eighteenth and nineteenth centuries, Lewis's book was reprinted numerous times under slightly different titles and with varying authors' names (Jane Lewis in one edition, Mrs. John Lewis in another, etc.). The futility of a search for information on a "real" Hannah Lewis corroborates the findings of captivity-narrative scholars Kathryn Derounian-Stodola and James Levernier, who write that "distinguishing historically verifiable first-person accounts from edited and fictionalized ones is often impossible owing to multiple authorial contributions, unclear publishing conditions and copyright, and generic overlap within and between works."[5] Lewis's narrative further typifies the genre in its narrator's emotional ambivalence toward her captors. Hannah's escape is made possible, in the book, only through her and her son's eventual success in leaving many of their old ways behind and gaining the trust of their captors. Recalling her final night among the "savages," she writes:

> My son now, in the true character of a savage, gave the Indian farewell whoop and departed. By my master he was recognized to be an adopted white prisoner, pleased with a savage life, but not as my son, who ten months before had visited his wigwam. As the savage dreamed of no design, I was permitted as usual to lie at night near the entrance of the wigwam, unguarded;—about twelve, three heavy blows with a hatchet upon a hallow tree nearby announced the approach of my son—it was the signal—from the hut I withdrew with cautious steps until I was received in the arms of my deliverer—seizing my hand, he hurried me to a thick swamp, which we penetrated for many miles—by his frequent excursions with the savages, he had obtained a knowledge of the wilderness, even to the white settlements, which proved of the greatest advantage to us—without meeting with a single savage to molest us, or any thing to retard our flight, in five days we safely arrived at the fort of St. Louis.[6]

---

5. Derounian-Stodola, Kathryn Zabelle, and James Arthur Levernier, *The Indian Captivity Narrative, 1550–1900* (New York: Twayne, 1993), 10.

6. Hannah Lewis, *The Narrative of the Captivity and Sufferings of Mrs. Hannah Lewis, and Her Three Children, Who Were Taken Prisoners by the Indians* (Boston: H. Trumbull, 1817), 24.

St. Louis, then, serves in Lewis's narrative primarily as a symbol for civilization—one too precarious still, in its proximity to the wilderness, to represent safety for families like the Lewises. The more refined pleasures of poetry made their debut in the wilderness outpost in 1821 with the publication of Angus Umphraville's *Missourian Lays and Other Western Ditties,* the first volume of poetry published west of the Mississippi. The book, dedicated to William Clark, comprised thirty-five poems, several of them celebrating the area's natural beauty, with emphasis on the splendor of the Missouri and Mississippi rivers. Umphraville made the most of his exotic material, imploring his reader not to "look for the genius of a Byron, a Moore, a Scott, a Campbell, or a Barlow, in 'the wood-notes wild' of Missouri." The poet continued: "With undissembled humility and diffidence, with fear and trembling, I offer this little volume to the reader; and as I have been severe on the hallucinations of contemporary writers, and not niggard in my encomia of their beauties, I may claim impartiality from the critic, whose indulgence I do not solicit."[7] The book is copiously annotated with references to Greek mythology, American Indian culture, the English Romantic poets, geography, and Christian doctrine, the net effect of which is to reveal Umphraville as a better scholar and prose writer than poet. In their pioneering assessment of Missouri's literary history, Elijah L. Jacobs and Forrest E. Wolverton wrote that "whatever may have been the popularity of the *Missourian Lays* at the time of its publication, any present renewal of interest in the volume can only be the result of historical rather than literary motives. After Umphraville's book, only a handful of volumes of verse were published in Missouri for more than a quarter of a century. Such poetry as was printed came to light chiefly in newspapers or other periodicals."[8]

Like Henry Shaw, a growing number of European immigrants made their way to the outpost town of St. Louis in the 1810s and 1820s. Intrepid, literate pioneers often wrote to those back in their homeland, offering details of New World living conditions, climate, and culture, and encouraging others to follow—or not. One such visitor, Gottfried Duden, an upper-class German, arrived in St. Louis in 1824. Enchanted both by the region's beauty and by its prospects as a site for land investment, Duden purchased approximately 270 acres of land in what is now Warren County. He eventually wrote and published his *Report on a Journey to the Western States of North America* (1829). Beautifully written in a personal letter format, Duden's *Report* contains detailed (if sometimes hyperbolic) descriptions of the St. Louis area's lush indigenous

---

7. Angus Umphraville, *Missourian Lays and Other Western Ditties* (St. Louis: Isaac N. Henry & Co., 1821), 5.

8. Elijah L. Jacobs and Forrest E. Wolverton, *Missouri Writers: A Literary History of Missouri, 1780–1955* (St. Louis: State Publishing Company, 1955), 19–20.

vegetation, rich soil, mineral deposits, strategic position on waterways, Indian population, and archaeological features such as the mounds on both sides of the Mississippi. His didactic postscript advises fellow Germans contemplating emigration to Missouri to consider the state as more than a temporary home:

> Any European who is tolerably informed and who comes to the Ohio, the Mississippi, or to the Missouri not entirely without money and does not disdain a normal occupation will find with some persistence an attractive sphere of activity, and a businessman who does not figure to see his hopes fulfilled in the first year should be advised to remain until he has become more familiar with the new things than is possible on a mere tour. He must have stayed among the Americans for at least six months. . . .
>
> It is very natural for an emigrant to have an inclination to keep, so to speak, one foot in his home country when he roams through foreign countries. Love for one's country can explain this sufficiently. However, he can afford this only in rare cases without considerable cost. As a rule, the emigrant will do well to renounce his home country for the first ten years and to occupy himself seriously with the thought of making America his home. That will give him that foundation that must be the prerequisite for many advantageous undertakings.[9]

Duden did not follow his own advice. He stayed in Missouri only four years; the disappointment of many of his fellow German emigrants, finding conditions in Missouri much harsher than those described in his romanticized report, led some among his countrymen to mock Missouri as "Duden's Eden." Nevertheless, the turmoil of the 1840s brought a huge wave of Germans to St. Louis in search of the religious and political freedoms denied at home. Among these "Forty-eighters" were many intellectuals and professionals, some of whom would, like novelist and entrepreneur Henry Boernstein, publish their impressions of the city in the time of its fastest growth.[10]

By the 1840s, more travelers were passing through St. Louis—by land or by riverboat—than ever before. Not surprisingly, the city became a staple feature of western travel writings. The most illustrious of the literary observers of this period was the novelist Charles Dickens, who had already, by the time of his 1842 American trip, garnered considerable fame in England. On the last leg of a six-month tour of America, Dickens and his wife traveled to St. Louis via

---

9. Gottfried Duden, *Report on a Journey to the Western States of North America and a Stay of Several Years along the Missouri (During the Years 1824, '25, '26, 1827)*, general ed. James W. Goodrich; ed. and trans. George H. Kellner, Elsa Nagel, Adolf E. Schroeder, and W. M. Senner (Columbia: State Historical Society of Missouri and University of Missouri Press, 1980), 260.

10. For more on Boernstein, see Kenneth H. Winn's essay in this volume.

Cincinnati and Louisville on two steamboats. Not unlike his countrywoman Mrs. Trollope, Dickens was forthright in his criticism of America—particularly, the crude mannerisms of Americans themselves. Although his *American Notes* presented, in most respects, a fairly neutral view of St. Louis, Dickens found himself unable to resist putting in a dig against the Mississippi River and its surrounding climate:

> What words shall describe the Mississippi, great father of rivers, who (praise be to Heaven!) has no young children like him? An enormous ditch, sometimes two or three miles wide, running liquid mud, six miles an hour: its strong and frothy current choked and obstructed everywhere by huge logs and whole forest trees: now twining themselves together in great rafts, from the interstices of which a sedgy, lazy foam works up, to float upon the water's top: now rolling past like monstrous bodies, their tangled roots showing like matted hair, now glancing singly by like giant leeches; and now writhing round and round in the vortex of some small whirlpool like wounded snakes. . . . [T]he weather very hot, mosquitoes penetrating into every crack and crevice of the boat, mud and slime on everything: nothing pleasant in its aspect.

Dickens, who spoke at the Mercantile Library during his stay in St. Louis, rested at the luxurious Planters' House Hotel, which he called "an excellent house," adding that "the proprietors have the most bountiful notions of providing creature comforts. Dining alone with my wife in our own room one day, I counted fourteen dishes on the table at once."[11]

Dickens, like most Englishmen, was disturbed by slavery and concerned for the problem it posed to the solidarity of the States. Five years after his visit, the Massachusetts Anti-Slavery Society published a book that presented the grim particulars of this institution as it existed in St. Louis: the *Narrative of William Wells Brown, a Fugitive Slave, Written by Himself.* Born into slavery in Kentucky, Brown was thirteen years old in 1827 when his master brought him to St. Louis, where he was subsequently loaned out to various businessmen, including Elijah Lovejoy, the publisher and later famed abolitionist who would become a martyr to the antislavery cause after his murder at the hands of an angry mob in nearby Alton, Illinois. The most poignant passages in Brown's narrative describe his experiences working for a slave trader identified only as Mr. Walker, who transported slaves from St. Louis to New Orleans for auction:

---

11. Charles Dickens, *American Notes for General Circulation* (London: Chapman and Hall, 1842), 76, 77.

William Wells Brown (wood engraving; frontispiece of Brown's last book, *My Southern Home* [Boston: A. G. Brown & Co., 1880]). The *Narrative*, Brown's 1847 abolitionist memoir, drew upon his experiences in and around St. Louis to depict the horrors of slavery in Missouri and elsewhere. He achieved international fame as an orator and writer. *Missouri Historical Society*

> When I learned the fact of my having been hired to a negro speculator, or a "soul driver," as they are generally called among slaves, no one can tell my emotions. Mr. Walker had offered a high price for me, as I afterwards learned, but I suppose my master was restrained from selling me by the fact that I was nearly a relative of his. On entering the service of Mr. Walker, I found that my opportunity of getting to a land of liberty was gone, at least for the time being. He had a gang of slaves in readiness to start for New Orleans, and in a few days we were on our journey. I am at a loss for language to express my feelings on that occasion . . .
>
> There was on the boat a large room on the lower deck, in which the slaves were kept, men and women, promiscuously—all chained two and two, and a strict watch kept that they did not get loose; for cases have occured in which slaves have got off their chains, and made their escape at landing-places, while the boats were taking in wood;—and with all our care we lost one woman who had been taken from her husband and children, and having no desire to live without them, in the agony of her soul jumped overboard, and drowned herself. She was not chained.

Though the journey south was a source of particular anguish to Missouri slaves, Brown put to rest any illusions that their lot was much better where they were: "Though slavery is thought, by some, to be mild in Missouri, when compared with the cotton, sugar and rice growing States, yet no part of our slaveholding country is more noted for barbarity of its inhabitants, than St. Louis."[12]

In 1849, two catastrophic local disasters—a cholera epidemic and a devastating downtown fire—added a more immediate note of distress to the mounting political tensions in St. Louis. The Great Fire, which started on the steamboat *White Cloud*, spread to two dozen additional steamboats moored at the levee—most of which were loaded with highly flammable materials like cotton, hemp, and tobacco. A stiff breeze on the evening of May 17, 1849, carried sparks into the city streets, ignited office buildings, and spread quickly. While the fire still burned, the *Daily Union* reported the particular toll taken on the city's primary venues for literary production, its newspapers: "Since it is not possible to list all the houses that were burned, we will name only the main buildings. They are the *Telegraph* office, the United States hotel, the *Reveille* office, the *Republican* office, the *People's Organ* office. Thus all the English newspaper presses in the city were destroyed except our own, which is located on the north side of Locust and fortunately escaped the danger."[13]

---

12. William Wells Brown, *Narrative of William Wells Brown, a Fugitive Slave, Written by Himself* (London: Charles Gilpin, 1849), 38, 26.
13. *St. Louis Daily Union*, May 18, 1849.

Although the 1849 fire was typical of the kind of disasters that befell a great number of American cities in the mid-nineteenth century, its impact on the downtown newspaper plants made it uniquely devastating to St. Louis's cultural climate. The greatest loss to the city's literary heritage came with the destruction of the office and plant of the *Reveille*. Founded by brothers Matthew and Joseph Field with Charles Keemle in 1844, the *Reveille* served not only as a source of local and regional news, but also as a significant literary pulpit, "prominent enough," according to Lorin Cuoco and William H. Gass, "to be the platform for Edgar Allan Poe's defense of his character against a libelous attack in the *New York Evening Mirror*."[14] The *Reveille* was a wellspring of new literature, including original poems, translations, songs, and tall tales of the genre that would come to be known as Southwestern humor. Joseph Field's "Death of Mike Fink" first appeared in the *Reveille,* as did poems and anecdotes by his brother, the actor Matthew Field, who also wrote under the pen name "Phazma" for the *New Orleans Picayune*. Their works, like the work of other *Reveille* writers, rendered a picturesque glimpse of a city still marked by the stamp of its frontier past. Matt Field's preservationist poem "The Chouteau House," about an eighteenth-century landmark destined to be overtaken by new urban development, is one of the most frequently reprinted regional poems of the era:

> Touch not a stone! An early pioneer
> Of Christian sway founded his dwelling here,
> Almost alone.
> Touch not a stone! Let the Great West command
> A hoary relic of the early land;
> That after generations may not say,
> "All went for gold in forefather's day,
> And of our infancy we nothing own."
> Touch not a stone!
> Touch not a stone! Let the old pile decay,
> A relic of the time, now pass'd away.
> Ye heirs, who own
> Lordly endowment of the ancient hall,
> Till the last rafter crumbles from the wall,
> And each old tree around the dwelling rots,
> Yield not your heritage for "building lots."
> Hold the old ruin for itself alone;

---

14. Lorin Cuoco and William H. Gass, eds., *Literary St. Louis: A Guide* (St. Louis: Missouri Historical Society Press, 2000), 42.

> Touch not a stone!
> Built by a foremost Western pioneer,
> It stood upon St. Louis bluff, to cheer
> New settlers on.
> Now o'er it tow'rs majestic spire and dome,
> And lowly seems the forest trader's home;
> All out of fashion, like a time-struck man,
> Last of his age, his kindred and clan,
> Lingering still, a stranger and alone;—
> Touch not a stone!
> Spare the old house! The ancient mansion spare,
> For ages still to front the market square;—
> That may be shown,
> How those old walls of good St. Louis rock,
> In native strength, shall bear against the shock
> Of centuries! There shall the curious see,
> When like a fable our story be,
> How the Star City of the West has grown!
> Touch not a stone![15]

Field's poem, while waxing nostalgic for the past, represented a departure from the folksy frontier lore that had dominated the pages of the *Reveille,* and that had helped spawn the careers of such St. Louis-based writers as Adolphus Hart, John Henton Carter, and Solomon Smith, as well as Mark Twain.[16] In his call for the cultivation of conscience and civic duty, as well as for preservation at the expense of profit, Field presaged the self-conscious urban sensibility that

---

15. Field's poem first appeared in J. C. Wild's *The Valley of the Mississippi Illustrated* (St. Louis: Chambers & Knapp, 1841; facsimile ed., St. Louis: Hawthorne Press, 1981), 32–33. One of the earliest pleas to the city to preserve its unique architectural heritage, the poem was not effective in saving the house, which was demolished in October 1841, three months after the poem's publication.

16. Adolphus M. Hart is best known as the author of the two-volume *History of the Valley of the Mississippi* (1852), which he self-published in St. Louis, and *Life in the Far West; or The Comical, Quizzical, and Tragical Adventures of a Hoosier* (c. 1850), which traces the adventures of an unsophisticated bumpkin in St. Louis; Twain's *Life on the Mississippi* (1883) has some of his most amusing commentary about St. Louis; and Solomon Smith, who operated the St. Louis Theater from 1837 to 1851, published several volumes of anecdotal memoirs, most notably *Theatrical Management in the West and South for Thirty Years* (1868). Carter wrote the humorous volume *The Log of Commodore Rollingpin: His Adventures Afloat and Ashore* (New York: G. W. Carleton, 1874). Excerpts from these works can be found in *Seeking St. Louis: Voices from a River City, 1670–2000,* ed. Lee Ann Sandweiss (St. Louis: Missouri Historical Society Press, 2000).

characterized St. Louis writing in the latter half of the nineteenth century. Field himself, however, would be unable to witness the change: lacking the resources to revive the *Reveille* after the fire, he returned to a career in the theater, meeting with little success and dying in 1856.

In the decade that followed the Great Fire, St. Louis rebuilt its downtown, prospered economically, and grew in population. Yet the city's status as slaveholding territory continued, in the minds of many, to cloud its future prospects. The best account of the eruption of this long-simmering issue into full-scale conflict comes from a forty-one-year-old William Tecumseh Sherman, who in April 1861 had just moved his family into a house on Locust Street between Tenth and Eleventh streets. Sherman, who had been stationed at Jefferson Barracks a decade before, had accepted the position of president of one of the city's many new street railroad companies and was eager to settle into a comfortable civilian life.[17]

Things changed quickly. On May 10, Sherman witnessed firsthand the riot that broke out as federal troops marched the men of a secession-leaning state militia from the militia's base at Camp Jackson (just east of today's intersection of Grand and Olive boulevards) through crowded city streets to the federal arsenal in South St. Louis. Standing on the street outside his home, in the company of his young son, Willie, as he would later write:

> I heard balls cutting the leaves above our heads, and saw several men and women running in all directions, some of whom were wounded. Of course there was a general stampede. Charles Ewing threw Willie to the ground and covered him with his body. Hunter ran behind the hill, and I also threw myself on the ground. The fire ran back from the head of the regiment toward its rear, and as I saw the men reloading their pieces, I jerked Willie up, ran back with him into a gulley which covered us, lay there until I saw that fire had ceased, and that the column was again moving on, when I took up Willie and started back for home round by way of Market Street. A woman and child were killed outright; two or three men were also killed and several others wounded. The great mass of people on that occasion were simply curious spectators, though men were sprinkled through the crowd calling out, "Hurrah for Jeff Davis!" and others were particularly abusive of the "damned Dutch." . . . A very few days after this event, May 14th, I received a dispatch from my brother Charles in Washington, telling me to come at once; that I had been appointed colonel

---

17. William Tecumseh Sherman, *Memoirs of Gen. W. T. Sherman, Written by Himself* (New York: Charles Webster & Co., 1892), 1:198–99.

"Terrible Tragedy at St. Louis, Mo." (wood engraving from *New York Illustrated News,* May 25, 1861). As federal troops broke up the state militia encampment at Camp Jackson and marched their prisoners downtown, businessman William T. Sherman took his young son outside to witness the commotion. Sherman would later provide the most riveting written account of the ensuing riot, which preceded his own call to Washington and his eventual placement as commander of the western armies of the Union. *Missouri Historical Society*

of the Thirteenth Regular Infantry, and that I was wanted in Washington immediately. Of course I could no longer defer action.[18]

Although numerous accounts exist of the Camp Jackson incident, Sherman's is perhaps the best: clear, detailed, and seen with a journalist's eye. Indeed, throughout the ensuing war, journalism and diary entries provided the most revealing and interesting perspectives on St. Louis's precarious place along the border of North and South. Among the most significant are the autobiography of former slave James Thomas, which captures the atmosphere in the city on the eve of war; Anthony Trollope's essay "Missouri," which described conditions in the city under martial law, including the horrendous living conditions at Benton Barracks; and Antoine Auguste Laugel's European perspective of the

18. Ibid., 1:202–3. Published in the last decade of his life, Sherman's memoirs, which include some of the correspondence he exchanged with Washington officials in the spring of 1861, reveal his reluctance at being pressed back into military service.

conditions in "this unfortunate State, ravaged in all directions and twice condemned to the horrors of invasion."[19] A contrasting—even jocular—account of the city during this tense period appears in Francis Grierson's essay "The Planters' House." Grierson captured the ambience and excitement of the city's premier hotel and meetinghouse during the Civil War, when Union top brass could be found smoking cigars in the same room with Confederate officers and politicians:

> The Planters' House! What did it not represent in the history of the Far West in the early days! To me it was St. Louis itself. This famous hotel typified life on the Mississippi, life on the prairies, life in the cotton-fields, life in the cosmopolitan city. It stood for wealth, fashion, adventure, ease, romance—all the dreams of the new life of the Great West. It was the one fixed point where people met to gossip, discuss politics, and talk business. It was the universal *rendezvous* for the Mississippi Valley. Here the North met the South, the East met the West. It looked like nothing else in the hotel world, but it always seemed to me it was intended more for pilots, river-captains, romantic explorers, far-seeing speculators, and daring gamblers.
>
> It was here the goatee type was seen in all its perfection. On some of the chins the tufts of hard, pointed hair gave a corkscrew look to the dark faces, which somehow harmonised well with the eternal quaffing of mint-juleps, sherry-cobblers, and gin cocktails.
>
> An hour spent in the Planters' House just before the great election was an experience never to be forgotten. All who did not want to shoot or be shot steered clear a course in some other direction, for here, in the bar and lobbies, were the true 'fire-eaters' to be met, and while some had already killed their man, others were looking for a man to kill.[20]

If the end of the Civil War reenergized the city's sluggish economy, it also seemed to many writers to mark the end of the time of the "fire-eaters" and their colorful company. In their place came such equally charismatic but generally more sober individuals as Joseph Pulitzer, who embodied the opportunity afforded resourceful immigrants in the increasingly heterogeneous city.

---

19. James Thomas, "St. Louis on the Eve of War," chap. 10 in *From Tennessee Slave to St. Louis Entrepreneur: The Autobiography of James Thomas*, ed. Loren Schweninger (Columbia: University of Missouri Press, 1984); Anthony Trollope, "Missouri," in *North America* (New York: Harper & Brothers, 1862); Auguste Laugel, *The United States during the War, 1861–1865* (New York: Ballière Brothers, 1866), quoted in Georges Joyaux, "Auguste Laugel Visits St. Louis, 1864," *Bulletin of the Missouri Historical Society* 14:1 (October 1957): 45.

20. Francis Grierson, "The Planters' House," in *The Valley of the Shadows: Reflections of the Lincoln Country, 1858–1863*, ed. Harold P. Simonson (New Haven: College and University Press, 1970).

Planters' House Hotel (from stereoscopic photograph by Hoelke and Benecke, c. 1870). Amid the tensions of the Civil War and martial law, this downtown landmark stood, in the words of writer Francis Grierson, "for wealth, fashion, adventure, ease, romance." *Missouri Historical Society*

Like others making their fortunes in the postbellum era—and like an earlier immigrant success story, Henry Shaw, who had begun to turn more concertedly to philanthropic projects just prior to the war—Pulitzer used his considerable clout and financial resources to enhance the quality of life in the city; in Pulitzer's case, the *Post-Dispatch* offered a platform from which to call for civic reform and improvement. Shaw himself wrote, though at a more modest level, complementing his philanthropic gestures with such small essays as his original *Guide* to the Garden, in which he stated: "Of all public resorts a scientific garden when properly kept, will be found to be, not only one of the most delightful mediums for intellectual gratification and amusement, but also one of the greatest of temporal blessings that can be enjoyed as a people."[21]

Shaw's more flamboyant contemporary Logan U. Reavis, who moved to St. Louis from Illinois in 1866, envisioned a very different, grandiose plan for the city. Reavis's primary contribution to the city proved to be not cash or botanical expertise but his rhetoric. A sometime teacher and publisher, he failed at his attempt to start a newspaper in St. Louis and, instead, channeled his energies into launching a dramatic campaign to make St. Louis the nation's capital. That plan found its expression in Reavis's tract *St. Louis, the Future Great City of the World: and Its Impending Triumph,* which circulated in various editions throughout the 1870s and 1880s.

Reavis's essay articulated St. Louis's prospects on a global scale, pointing out its geographical superiority to other American cities and comparing its position within a great interior river valley (the same valley from which Charles Dickens had fled in horror) to the situation of Thebes, Memphis, and Rome:

> In the Mississippi Valley which is still new in its development, there are already many large and flourishing cities, each expecting, in the future, to be greater than any one of the others. First among these stand Chicago, Cincinnati, St. Louis, and New Orleans—four cities destined, at no distant day, to surpass in wealth and population the four cities of the Atlantic seaboard—Boston, New York, Philadelphia, and Baltimore. . . . That Chicago, Cincinnati, St. Louis, and New Orleans have each many natural advantages, there can be no question. There is, however, this difference: the area of surrounding country, capable of ministering to the wants of the people and supplying the trade of a city, is broken, in the case of New Orleans, by the Gulf of Mexico, Lake Ponchartrain and by regions of swamps. In the case of Chicago, it is diminished one-third by Lake Michigan; while Cincinnati and St. Louis both have around them unbroken and uninterrupted areas of rich and productive lands, each capable of sustaining a large population. But if it be asked to which of these cities belong the greatest advantages, must we not answer, it is the one nearest the center of the productive power of the continent, and especially to the natural wealth of the Valley of the Mississippi. Most certainly, for there will grow up the human power. And is not this center St. Louis? We have only to appeal to facts to establish the superior natural advantages of St. Louis over any other city on the continent.[22]

---

21. Henry Shaw, "A Guide to the Missouri Botanical Garden, St. Louis," *Missouri Botanical Garden Bulletin* 31:7 (September 1943): 136–45.

22. Logan Uriah Reavis, *St. Louis, the Future Great City of the World: and Its Impending Triumph* (St. Louis: G. A. Pierrot, 1881), 14–15.

Logan Uriah Reavis (lithograph by Forbes and Co., c. 1875). Reavis, convinced that St. Louis was "the future great city of the world," produced a body of literary hyperbole that has since been unmatched in this or perhaps any other American city. *Missouri Historical Society*

Reavis actively promoted the capital relocation scheme in the United States and England, long after the most fervent St. Louis boosters abandoned the plan. After the 1880 census, when St. Louis was confronted with the reality—made more acute by the inflated figures of the 1870 census—that it had fallen behind Chicago, Reavis became the target of public scorn and ridicule—"not because his plan lacked merit," as James Neal Primm argues, "but because it had failed."[23] Nevertheless, Reavis's tract is a pivotal document in nineteenth-

23. James Neal Primm, *Lion of the Valley, 1764–1980*, 3d ed. (St. Louis: Missouri Historical Society Press: 1998), 275.

century St. Louis literature, for it weds the early hyperbole of frontier-era writing with a new sense of civic identity, one that builds on St. Louis's place at the heart of a global urban network. Like a prophet howling in the wilderness, his message falling on essentially deaf ears, Reavis was either too advanced or too crazy for his time. Yet he proved to be the forerunner of a man of letters and civic activist whom St. Louis could not ignore: William Marion Reedy.

Born in North St. Louis's Kerry Patch neighborhood in 1862 to an Irish policeman and his wife, Reedy attended Christian Brothers Academy and St. Louis University and embarked on his career as a journalist at a time when Reavis's star was in decline. Reedy gained his first taste of literary life in 1882 when, as a cub reporter for the *Missouri Republican,* he was assigned to interview a visiting Oscar Wilde. The flamboyant and controversial Wilde, under constant and vitriolic attack by the American press, impressed Reedy with his candor and courage. Critical of the priggishness and conventionalism of St. Louis journalists, Reedy soon gravitated to magazine writing and publishing, where he enjoyed a wider editorial berth as well as the opportunity to promote writers of great promise.

Reedy's vehicle for entering publishing was a floundering St. Louis magazine called *The Mirror,* which he took over in 1893 and subsequently turned into one of the country's outstanding periodicals. As editor, Reedy consciously positioned *The Mirror* in opposition to the genteel and, in his estimation, irrelevant literary magazines that dominated St. Louis's literary scene in the 1890s.[24] A showcase for emerging literary talent, *The Mirror* published the early work of such future luminaries as Theodore Dreiser (who wrote briefly for both the *Globe-Democrat* and the *Republican*), Kate Chopin, Ambrose Bierce, Sara Teasdale, Carl Sandburg, Fanny Hurst, Edgar Lee Masters, and Zoe Akins. Reedy himself became an influential, nationally known literary critic and afficionado of naturalism who championed the novels of Stephen Crane, Thomas Hardy, and Dreiser.

True to Reedy's original mission, *The Mirror* was also the source of pointed, stimulating essays and articles on local events and controversies. In the waning weeks of the nineteenth century, Reedy published an editorial that read like a more confrontational counterpart of the address given by Amos Stoddard in 1804. In characteristically bold fashion, Reedy titled his essay "What's the Matter with St. Louis?—An Attempt at a Diagnosis of a Generally Observed Ailment in the Fourth City of the Union." In this full-frontal attack, Reedy

---

24. Max Putzel, *The Man in the Mirror: William Marion Reedy and His Magazine* (Cambridge: Harvard University Press, 1963), 53.

lambasted middle-class St. Louisans for their lethargy and lack of civic pride, and accused the wealthy power brokers of greed and exploitation, pointedly stating: "St. Louis is belittled by its own people. There is no local evidence of any general love for the city. There is no indication of any feeling working to the end of making this a better place to live in. With a few rare exceptions, St. Louis rich men have never given of their substance for any great public institution or improvement that promised them no immediate cash return. What man has given anything to the city for the use of all the people since Henry Shaw died?"[25]

Beyond its general call for a new civic-mindedness, the specific intent of Reedy's essay was to rouse St. Louis to commit to hosting a World's Fair in commemoration of the centennial of the Louisiana Purchase.[26] Reedy made the connection clear in his closing words:

> The matter with St. Louis, then, is too much mere matter, too little mind. The people who predominate lack vision. The city is devoid of any community of affectionate desire to make the city great and beautiful as an expression of something besides mere sordid solidity in a financial way. St. Louisans should love St. Louis as Londoners love London, Parisians love Paris, New Yorkers love New York, Chicagoans love Chicago. They should unite in a desire that the world should know their city's merits as a home of people with hearts responsive to beauty and all exalted things.
> . . . A World's Fair will awaken this city's anaemic soul. It will breathe a soul into beings who now seem not to know what soul means. It will awaken the dormant spirit in many, as the sight of the Acropolis strikes reverence from the silliest "Cookie." We need it that we may not attain a piteous and contemptible distinction as a village bloated by sedentary satisfaction into a metropolis of Boeotia. St. Louis needs an awakening, such as described, as much as Havana needed sanitation. Let us work for it.[27]

Just as Amos Stoddard had almost a century earlier, Reedy challenged St. Louisans to own up to the responsibilities of citizenship and to dare to invest in a dream. Despite the fact that it infuriated many St. Louisans, despite its relative insignificance as a piece of literature, Reedy's indictment worked. According to his biographer, Max Putzel, "[T]he response filled bushel baskets; the magazine sold out at once. When the article was reprinted in a new pamphlet series, the

---

25. William Marion Reedy, "What's the Matter with St. Louis?" in Lee Ann Sandweiss, ed., *Seeking St. Louis: Voices from a River City, 1670–2000* (St. Louis: Missouri Historical Society Press, 2000), 402–8.
26. Primm, *Lion of the Valley*, 374–76.
27. Reedy, "What's the Matter with St. Louis?" 408.

pamphlet was sold out in turn. Afterward, Reedy dated the success of the St. Louis World's Fair and the civic reform movement that paved the way for it to the day his article appeared in November 1899."[28]

St. Louisans living amongst the social and environmental uncertainties of the early twenty-first century can still hear, in the rhetoric of today's journalists, pundits, and civic leaders, the echo of Stoddard's and Reedy's challenges. In the decades that followed Stoddard's address to his new countrymen, the call for civic betterment—led by true believers as diverse as the Field brothers, Joseph Pulitzer, Henry Shaw, and Logan U. Reavis—formed an important strain within the city's rich literary heritage. In the last years of the century, as concern began to overshadow optimism, it was a man of letters, William Reedy, who proved as effective as politicians or business leaders in mobilizing St. Louisans to purify their water, pave impassable streets, clear trees in Forest Park, and put on a phantasmagoric spectacle that they have never forgotten. At the dawn of this new millennium, we have yet to see who will pen the clearest such message today—and how St. Louisans will respond to it.

28. Putzel, *Man in the Mirror*, 94.

Contributors

ROBERT R. ARCHIBALD is President of the Missouri Historical Society. A board member of the American Association of Museums and past president of the American Association for State and Local History, he has written and lectured extensively on new directions in public history. Archibald is the author of *A Place to Remember: Using History to Build Community* (Walnut Creek, Calif.: AltaMira Press, 1999).

LOUIS GERTEIS is Professor of History and Chair of the Department of History at the University of Missouri–St. Louis. He is the author of three books, most recently *Civil War St. Louis* (Lawrence: University Press of Kansas, 2001). Among his articles is "St. Louis Theatre in the Age of the Original Jim Crow," *Gateway Heritage* (Spring 1995).

ANTONIO F. HOLLAND is Professor of History and Chairman of the Division of Social and Behavioral Sciences at Lincoln University in Jefferson City, Missouri. He is coauthor of *The Soldiers' Dream Continued: A Pictorial History of Lincoln University* (Jefferson City: Lincoln University, 1991) and *Missouri's Black Heritage* (Columbia: University of Missouri Press, 1980; rev. ed., 1993).

WALTER D. KAMPHOEFNER, born and educated in Missouri, teaches immigration history at Texas A & M University. His extensive publications on the social and political history of German immigration include a monograph, *The Westfalians: From Germany to Missouri* (Princeton: Princeton University Press, 1987), and a coedited letter collection, *News from the Land of Freedom: German Immigrants Write Home* (Ithaca: Cornell University Press, 1991). His most recent work is a nationwide anthology of German-American Civil War letters, scheduled to appear in Germany in 2002 and subsequently in English translation.

MICHAEL LONG is an adjunct faculty member of Webster University in St. Louis, and a Research Associate at the Missouri Botanical Garden. He is at work on a biography of the physician and scientist George Engelmann, the man who helped Henry Shaw to found his botanical garden.

JAMES NEAL PRIMM, Curators' Professor of History Emeritus at the University of Missouri–St. Louis, is a native of Edina, Missouri. Among his books are three on Missouri topics: *Economic Policy in the Development of a Western State, Missouri, 1820–1860* (Cambridge: Harvard University Press, 1954); *Lion of the Valley: St. Louis, Missouri, 1764–1980* (1981; 3d ed., rev., St. Louis: Missouri Historical Society Press, 1998); and *A Foregone Conclusion: The Founding of the St. Louis Federal Reserve Bank* (St. Louis: Federal Reserve Bank of St. Louis, 1992).

PETER H. RAVEN has been the Director of the Missouri Botanical Garden and Engelmann Professor of Botany at Washington University in St. Louis since 1971. One of the world's leading authorities on plant systematics and evolution, he has published more than 480 books and papers in the fields of taxonomy, population biology, biogeography, reproductive biology, ethnobotany, and conservation biology. Recipient of the National Medical of Science, he champions research around the world to preserve endangered plants and is a leading advocate for conservation and a sustainable environment.

WILLIAM J. REESE is Professor of Educational Policy Studies, History, and European Studies at the University of Wisconsin–Madison. Former editor of the *History of Education Quarterly,* he is the author of *The Origins of the American High School* (New Haven: Yale University Press, 1995).

ERIC SANDWEISS is Carmony Associate Professor of History at Indiana University and the editor of the *Indiana Magazine of History.* The author of *St. Louis: The Evolution of an American Urban Landscape* (Philadelphia: Temple University Press, 2001), he served for ten years as Director of Research at the Missouri Historical Society.

LEE ANN SANDWEISS is the former Director of Publications at the Missouri Historical Society. In addition to the numerous books on St. Louis history which she helped to develop in her work there, she is the author of *St. Louis Architecture for Kids* (St. Louis: Missouri Historical Society Press, 2001) and editor of *Seeking St. Louis: Writings from a River City, 1670–2000* (St. Louis: Missouri Historical Society Press, 2000).

KENNETH H. WINN is the State Archivist of Missouri. He is the author of *Exiles in a Land of Liberty: Mormons in America, 1830–1846* (Chapel Hill: University of North Carolina Press, 1989) and coeditor of the *Dictionary of Missouri Biography* (Columbia: University of Missouri Press, 1999).

Index

Page numbers in italics refer to illustrations.

Abolitionism, 32, 60, 211, 225
Academy of Science of St. Louis, 32; acquisitions by, 155–57; decline of, 163–66
Adams, John Quincy, 28
Address to the Friends of Equal Rights (Equal Rights League), 70
"Address to the People of Upper Louisiana" (Stoddard), 221
Adler, Jeffrey S., 134
African Americans, 37, 66, 92–94; depicted in art, 211–15, *214;* depicted in plays, 200–201, 216; discrimination against, 73, 77–78; education for, 46, 169, 179, 186; in politics, 73–74; population in St. Louis, 34, 84, 89; as portrayed in blackface performances, 193–94, 197–99, 201–4; vote for, 48, 69–71. *See also* Free blacks; Slaves
Agassiz, Louis, 141, 148, 152
Agrarian empire, based in St. Louis, 19, 50
Agrarianism, 36–37
Agrarian reform, 26
Agriculture: products of, 112, 121–22; prosperity of Missouri's, 34–35; in St. Louis economy, 125
Aiken, George L., 215, 217
Allen, Andrew, 193–94
Allen, Ann Russell, 115–16
Allen, Gerard B., 112, 129, 132–34
Allen, Thomas, 112; and railroads, 115–19, 129, 131; as Yankee, 132, 134
American Association for the Advancement of Science, 148
American Colonization Society, 65
American elite, 24
American Fur Company, 153–54
American Missionary Association, 67

American Notes (Dickens), 31, 225
"American System," 28
Ames, Henry, 132
Anarchists, 94
Anderson, Galusha, 90
Anglo-Americans, relations with immigrants, 86, 97–98
Anticlericalism, 41–42
Anzeiger des Westens (newspaper), 82
Arbeiter Zeitung (newspaper), 94
Arikara Indians, 107
Art: genre, 193, 210–15; St. Louis's prominence in, 217
Ashley, William H., 107–8
Astor, John Jacob, 107
Atchison, David Rice, 36–37
Audience: in public masquerade, 209–10; role in theater productions, 192–93, 197, 215
Audubon, John J., 151, 153

"Backside Albany" (Hawkins), 193–94
Bacon, Henry, 123, 132, 134
Baker, Jesse, 111
Ball, W. C., 74
Baltimore, Maryland, 114, 130–31
Bank of Illinois, 109–11
Bank of Missouri, 105, 109
Bank of St. Louis, 125
Bank of the State of Missouri, *110,* 124–25
Bank of the United States, 26, 28, 109
Banks: free banking vs. limited, 124–25; investing in railroads, 117, 119, 123; state vs. private, 108–11, 125
Bannon, John B., 90
Barboa, Joseph, 53
Bargaining for a Horse (Mount), 211

241

Index

Barnett, George I., 5
Barry, James, 115–16
Barth, Gunther, 9
Barth, Robert, 128, 133
Bartholdt, Richard, 92
Barton, David, 28
Barton, Joshua, 28
Baseball, Chicago-St. Louis rivalry in, 134–35
Bates, Edward, 26–29, *27,* 43–44
Bates, Frederick, 27
Bates, John, 216*n*33
Beatty, Barbara, 177
Beckwourth, Jim, 63
Beecher, Lyman, 215
Belcher, Charles and William, 124, 134
Belcher, Wyatt, 131–34
Belcher Sugar Refining Company, 123–24
Benoist, Louis A., 111, 123, 133
Benton, Thomas Hart, 22–26, *23,* 37, 39, 109; constituency of, 34–35, 42; political career of, 28–29, 35–36, 43; on railroads, 114, 116
Benton Barracks, 66, 127
Bernhardi, Johann Jakob, 159
Berry, John, 95
Biddle, Thomas, 109
Bingham, George Caleb, 193, 212–14
Bischoff, G., 151
Black codes, 51–52, 55–58
Blackfeet Indians, 107
"Black Republicans," 43, 45
Blacks. *See* African Americans
Blair, Francis P., Jr., 37, *38;* opposition to Reconstruction, 47–48; opposition to secession, 39–40, 127; politics of, 42–44, 93
Blair, Montgomery, 37, 39, 43–44
Blakeley, Tom, 201
Blood, Sullivan, 132, 134
Blow, Henry Taylor, 132, 133
Blow, Peter, 45
Blow, Susan, 97, 169, *176,* 177–79, 187
Boatmen's Savings Association, 123
Boernstein, Henry, 41–42, 94, 224
Bogy, Louis V., 133
"Bone Squash Diavolo" (Rice), 205–9
Booneslick region, 34
Border states, immigrants to St. Louis from, 103–4
Boston, Massachusetts, 121, 130–31
Botany, 166; in development of Shaw's garden, 157; Engelmann's work on, 146, 149–52, 159, 161; Shaw's interest in, 5–7; specimen collectors, 150–52
Brant, Joshua, 132

Braun, Alexander, 146
Bridge, Hudson, 111–12, 128, 132, 134; and railroads, 119, 128, 131
Bridges, across Mississippi River, 49. *See also* Eads Bridge
British immigrants, 83
Brokmeyer, Henry, 97
Brooklyn, New York, 99, 103, 125
Brooks, William P., 68
Brown, B. Gratz, 37; politics of, 42–43, 47–48, 93; postwar, 47, 50
Brown, Benjamin B., 141, 149
Brown, Charles, 60
Brown, William Wells, 59, 225–27, *226*
Bruns, Jetta, 86
Bryant, George, 74
Budd, George, 111
Burgess, Albert, 74
Business, 199; and banking crises, 110–12, 123; effects of Civil War on, 127–29; number and diversity of, 75, 104; and science, 138, 153; support for railroads, 115–20. *See also* Trade
Butler, Ed, 74, 91
Butler, Jim, 74

Calhoun, John C., 35, 36
California, and St. Louis banks, 123, 124
Camden, Peter, 87
Campbell, Robert, 108, 133
Camp Jackson incident, 40, 89, 230–32, *231*
Carondelet, annexation of, 130
Carpenter, Charles J., 140–41
Carstang, Effie, 4
Carter, John Henton, 229
"Cascades, The" (Joplin), 76
Catholics, 62, 91; cathedral of, 104–5; and German immigrants, 92, 94; limited discrimination against in St. Louis, 87, 89
Central Clique, of Democratic party, 34, 36
Channing, William Henry, 31
Charless, Joseph, 80
Charleville, Louis, 65
Chase, Salmon, 44
Chicago: compared to St. Louis, 49–50, 104, 125; effects of Civil War on, 126–27; immigrants in, 83–84, 99; rail connections to, 115, 118, 164; St. Louis economic rivalry with, 49–50, 125–26, 130, 134; St. Louis rivalry with, 130–32, 134–35, 164
Cholera epidemic (1849), 121, 152
Chouteau, Auguste, 21–22, 55, 80, 126; narrative of, 2, 219–21

Chouteau, Charles P., 123–24, 133, *154;* and railroads, 119, 132; support for science, 138, 153–56, 163
Chouteau, Marie Thérèse, 115*n*36
Chouteau, Pierre, 21–22, 56, 80, 112
Chouteau, Pierre, Jr., 107–9, 132, 149, 153; influence of, 110, 133
"Chouteau House, The" (Field), 228–29
Christy, Andrew, 119, 123
Churches: activism of, 62, 69, 72; Bohemian, 94; construction of cathedral, 104–5; of free blacks, 62, 63; German, 82; newspapers of, 94
Cincinnati, Ohio, 113–14, 126, 130–31
Cincinnati Commercial Agency, St. Louis branch, 109–10
Cities, 83; culture of, 8–10, 218; growth of, 9, 73–74, 125; schools in, 167–69
Citizenship: free blacks denied, 60; for immigrants, 87, 95
City Bank (St. Louis), 125
Civil rights: black codes vs., 51–52, 55–58; for blacks, 37, 60, 68–69, 71–73
Civil War, 32, 47, 49; blacks fighting in, 66, 69–70; effects of, 84, 120, 163; effects on St. Louis economy, 103, 127–28, 134; immigrants in, 89–90, 94; Missouri's participation in, 40, 89; reports of St. Louis during, 231–32; Union support in St. Louis, 6, 20, 35
Clamorgan, Cyprian, 65
Clamorgan, Jacques, 52–53
Clark, Charles C., 75
Clark, Crittenden, 75
Clark, Meriwether Lewis, 145, 149
Clark, Peter, 74
Clark, William, 25, *137;* museum of, 136, 138, 145; as territorial governor, 22, 27
Clarke, Enos, 70
Clarke, James Freeman, 31
Clay, Henry, 28
Clemens, James, 109
"Coal Black Rose" (Blakeley), 201–2
Code Noir, 51–52
Cole, R. H., 74
Collier, George, 110, 116, 132
Colonization, black, 44, 65
"Colored aristocracy," 65
Colored Relief Board, 72
Commerce of the Prairies (Gregg), 152
Confederacy, 47, 120; supporters of, 89–90; treatment of supporters of, 46, 48, 50
Congress, U.S., 28, 36, 117, 145

Constitution, Missouri, 27–28, 35, 46, 50; black rights under, 69–70, 73
Corbin, Abel R., 114
Corruption, 48, 128–29
Cottmann, E. T., 74
Cotton Belt Railroad, 130
Cotton Exchange, 129
Cotton trade, 129–30
County Election, The (Bingham), 214
Cowell, Joe, 201
Creoles, 79–80, 87, 132, 134
Cristophe, Henri, 53
Crittenden, Thomas T., 73
Crow, Wayman, 112, 133
Cruzat, Francisco, 55
Culture: definition of, 8–9; German, 42, 174; institutions of, 32; of St. Louis, 14, 34–35
Cuoco, Lorin, 228
Currency: hard money vs. paper, 25–26; issued by states, 106, 109–11, 125

Daniels, George H., 143
Deas, Charles, 193, 212
"Death of Mike Fink" (Field), 228
DeBar, Ben, 216
Decaisne, Joseph, 151
Democratic party, 46; and blacks, 74–75; Central Clique of, 34, 36; ethnocultural background of, 86–87; and German immigrants, 42, 88–89; on slavery issue, 37–39; split in, 36, 87
Derounian-Stodola, Kathryn, 222
Dessalines, Jean-Jacques, 53
Des Ursins, 51
Dickens, Charles, 31, 224–25
Dickson, Moses, 69, 72, 74
Diseases, 105, 121, 152
Distilleries/breweries, 126, 129
Divoll, Ira, 168, 170
Doniphan, Alexander, 36–37
Drake, Charles, 44–48
Drake, Mary Ella Taylor Blow, 45
Dreiser, Theodore, 13, 236
DuBourg, Louis, 104
Duden, Gottfried, 82, 141, 225–26

Eads, James, 48–49, 85, 132–34, 164
Eads Bridge, 49, 131
Easton, Rufus, 27
Economy, 28, 34, 124; blacks in, 68, 75; effects of slavery and sectionalism on, 37–39, 126–27, 134; immigrants' influence on, 99, 112; importance of St. Louis in national,

19–20; Missouri's, 50, 117; problems of, 105–6, 123; recovery of, 106, 112; of rural Missouri vs. St. Louis, 34–35; St. Louis rivalry with Chicago in, 49–50, 125–26, 130, 134; wealth of St. Louis, 103, 106–8. *See also* Financial panics and depressions
Education: benefits of, 168–69, 173–74; for blacks, 58, 62–63, 66–68, 70–71, 73, 77–78, 186; controversy over vocational training in, 181–86, *185;* criticism of methods in, 172–73, 175; as democratizing, 175, 179; development of St. Louis school system, 170–72; in German, 86, 95–97, *96;* innovations in, 167–68, 174–75; kindergartens in, 73, 175–80, *178;* moral goals of, 173–74, 178; new institutions of, 153; pedagogical methods in, 180–81, 185, 187; prominence of St. Louis in, 168–70; public vs. private schools, 167, 168, 171–72; support for, 32, 46; of women, 153*n*54
Eighth Street Baptist Church, 72
Elections, violence around, 88
Eliot, William Greenleaf, 29–33, *30,* 144, 215–16
Elleardsville, 78
Emancipation, of slaves, 65, 67, 92; methods of, 52, 60–62; Radical movement for, 45–47
Emerson, Irene, 115*n*36
Emerson, Ralph Waldo, 31
Engelmann, George, 138, 141, *142,* 144, 146–53, 161; and Academy of Science of St. Louis, 155, 165; and Shaw, 7, 157–61
Equal rights, blacks fighting for, 70–71
Equal Rights League, Missouri, 69–71
Ethnocultural polarization, 86–88
Europe, 5, 99, 150, 167; reasons for emigration from, 40–42, 82. *See also* Germany, revolutions in
Evangelicals, German-language publications by, 94
Exchange Bank, 125
Exodusters, 71–72

Faherty, William Barnaby, 3, 5
Falkenhainer, Heinrich, 121
Falkenhainer, Melchior, 121
Farmer, Walter M., 74, 77
Farmers, 211; Benton's support for, 22, 25–26, 37; German immigrants as, 40, 43
Farmers Nooning (Mount), 211
Fendler, August, 152
Ferguson, H. S., 75
Field, Joseph M., 217, 228

Field, Matthew, 228–30
Fields, W. H., 75
Filley, Chauncey I., 72–74, 133
Filley, Oliver and Giles, 112, 132, 134
Fillmore, Millard, 43
Financial panics and depressions, 169; of 1819–1823, 107, 191; of 1837, 24, 45, 149; of 1839, 114; of 1857, 119, 124; of 1873, 68
Finkelburg, Gustav, 92–93
Fire of 1849, 121, 227–28, 230
First African Baptist Church, 62
First Missouri Infantry of African Descent, 66
First Regiment of Missouri Colored Infantry, 66
First Ward, St. Louis, 85, 92
Fletcher, Thomas, 46
Flint, Timothy, 31, 62
Flora of North America (Gray), 150
Food processing, in St. Louis economy, 126, 129
Forchet, Jeanette, 52, *54*
Forest Rose, The (Woodworth), 193, *196,* 197–200, 217
"Forty-eighters," 41–42, 224
Fossils: in early museums, 136, 139–40; private collections of, 154–55
"Founding of St. Louis" (Cameron), *220*
France: influence of colonialism, 51–52; race relations under, 51–53, 55; trade rights granted by, 20–21
Franklin Society, 144
Free blacks: education for, 62–63, 66–67; institutions of, 51, 63; jobs of, 63–65; population in Missouri, 55, 59, 68; population in St. Louis, 52, 59, 125; property of, 52–53, 65; in St. Louis, 52–53, 59
Freedmen's Bank, 68
Freedmen's Bureau, 67
Freedmen's Orphan Home, 68
Free-Soil Democrats, 39, 87–89
Fremont, John C., 116, 146, 151
French: elite, 19, 21–22, 24, 87, 131; as linguistic minority, 79–80; number of immigrants, 83
Fresenius, Georg, 146
Froebel, Friedrich, 177–80
Fuller, Margaret, 29
Fur trade, 21, 24, 103, 107–8, 112

Gamble, Hamilton, 44–46
Garrison, Daniel R., 128
Gass, William H., 228

Gaty, Samuel, 119, 132–33
Genre art, 193, 210–15
German, education in, 95–97, 172, 174
German Evangelical congregation, 82
German immigrants, 226; culture of, 41–42, *98;* discrimination against, 35, 88–89, 91; influence in St. Louis, 91, 103–4, 133, 174; number of, 34, 83–85, 113, 121; opposition to slavery, 36, 39; in politics, 42–44, 47, 87, 92; relations with Irish, 85, 88–89, 91, 95–96; in St. Louis, 35, 80–83; support for the Union, 40, 89–91
Germany, revolutions in, 40–41, 83, 121, 224
Geyer, Henry, 36, 80
Geyer, Karl, 146, 150–51
Giddings, Salmon, 31
Gilpin, William, 132
Goethe, Johann Wolfgang von, 141
Gold, St. Louis banks handling, 123
Gould, Jay, 49, 129
Graham, Richard, 115*n*36
Grant, Ulysses, 47–48, 71
Gratiot Street prison, 163
Gray, Asa, 149–52, 157–58
Greeley, Carlos, 112, 124, 131–32, 134
Greeley, Horace, 48
Green Mountain Boy, The (Paulding), 199–201
Gregg, Josiah, 152
Grierson, Francis, 232
Guerrilla activity, 46

Hackett, James Henry, 197, *198,* 202–3, 211, 217
Haiti, black rebellion in, 53, 55
Hall, James, 153
Handy, W. C., 77
Hannibal and St. Joseph Railroad, 114, 116–17, 120
Hard money, vs. paper, 25–26
Harney, U. S., 115*n*36
Harney, William S., 39
Harris, William Torrey, 97, 169–70, *170;* and addition of kindergartens, 177–79; on effects of education, 183–84; on goals of education, 173–74; influence of, 174–77, 186–87; opposition to vocational education, 181–83
Harrison, James, 124, 132–33; and railroads, 116–19, 128; pledging money to guarantee banks, 123–24
Harrison, William Henry, 21
Hart, Adolphus, 229
Hawes, Harry B., 75

Hawkins, Micah, 193, 211
Hayden, Ferdinand V., 152–54
Hazlett, Sarah, 65
Hegel, Georg Wilhelm Friedrich, 97, 170
Hemp trade, 112
Henderson, Madison, 60
Hendrickson, Walter B., 148
Herschel, John, 146
Hill, George Handel "Yankee," 197, 199–201, 217
Hodgen School, Manual Training Room, *185*
Holbrook, John E., 147
Home-rule charter, 130
Hooker, William, 5, 157, 159
Hosmer, Harriet, 153*n*54
Housing, 75, 78, 104
Howard, Cordelia, 215–16
Howard, George C., 215–16
Hubbard, Jesse, 62
Hunt, Anne, 115*n*36
"Hunters of Kentucky, The" (Woodworth), 193–94
Hunting expeditions, from St. Louis, 19
Hutchinson, Asa, 216

Illinois, 82, 109; railroads of, 114, 117, 132; relation to St. Louis, 89, 119. *See also* Chicago
Illinois and St. Louis Bridge Company, 132
Illinois River, 113
Immigrants, 8, 36, 95, 113, 125, 150; influence of, 84, 103–4, 112, 232–33; relations among, 88–89; sources of, 99, 112; travel narratives and journals of, 225–26; vs. natives, 41, 86–89, 211. *See also* German immigrants; Irish immigrants
Immigration, 20, 79, 82; and growth of St. Louis, 19, 33; obstacles to, 37, 84; peak of, 87–88
Indiana, 119
Industry. *See* Manufacturing, in St. Louis economy
Insurance companies, 111
International Order of Twelve Knights and Daughters of Tabor, 69
Irish, 93–95
Irish immigrants, 35, 80, 88, 90; influence of, 91, 103–4, 133; number of, 83–85, 113; politics of, 43, 87; relations with Germans, 85, 88–89, 91
Italian immigrants, 99

Jackson, Andrew, 24–26, 28, 194–95
Jackson, Claiborne Fox, 40
Jackson, David, 107
Jacksonian democracy, 26, 35
Jacobs, Elijah L., 223
January, Derrick A., 112, 123–24, 127, 133
Jefferson, Joseph, 192–93
Jefferson Barracks, 127
Jefferson Club, 75
Jews, 94, 96, 99
"Jim Crow" material, 201–2, 205, *206*
Jobs, 126; for blacks, 53, 63–65, 68, 75; and education, 181–84; for Germans vs. Irish, 93–94; of slaves, 56. *See also* Labor
Johns, Elizabeth, 210–11
Johnson, Eastman, 215
Johnson, William, 65
Jolly Flatboatmen, The (Bingham), 212
Jolly Flatboatmen in Port, The (Bingham), 212–14, *214*
Jones, Michael, 32
Joplin, Scott, *76*, 77
Journal of Speculative Philosophy, 97
Judiciary, 47, 60

Kansas, blacks fleeing to, 71–72
Kansas City, Missouri, 74
Kansas Pacific Railroad, 131–32
Kayser, Henry, 91
Keckley, Elizabeth, 63, *64*
Keemle, Charles, 228
Kelly, Joseph, 90
Kennett, Luther M., 116, 123, 131, 133
Kentuckian, The (Paulding), 197, 202–4, 217
Kentuckian character, *198*, 202–4; in art, 212, *213*; in theater, 193–95
Kentucky, 133; migration to St. Louis from, 34, 89, 103–4, 106
Kiel, Henry, 92–93
Kindergarten. *See* Education
King, Edward, 98
King, Henry, 140–41, 148–49
Kirkland, Edward C., 120
Knights of Liberty, 69
Know-Nothing party, 42–44. *See also* Nativism
Koch, Albert, 139–40, 156–57
Krum, John M., 60

Labor: commodification of, 209; criticism of movement, 6; free vs. slave, 33–34, 36–37, 39, 104
Laclede, Pierre, 20–21, 220–21
Land, for settlers, 25, 28

Land grants: for railroads, 117, 119–20; from Spain, 21–22, 24
Lane, William Carr, 80
Langston, John M., 70
Languages, 172, 174. *See also* German, education in; Linguistic groups
Laugel, Antoine Auguste, 231–32
Lawless, Luke, 60
Lead mining, 51, 53
Leduc, Marie P., 145, 149
Legislature, Missouri, 28, 45, 47; blacks' treatment under, 39, 70, 73; lack of support for science, 145, 164; and railroads, 114–16, 128, 131–32; regulation of banks by, 109, 111, 125
Leighton, George, 127
Lemp, William, 126
Levernier, James, 222
Levine, Lawrence, 192
Lewis, Hannah, 222–23
Lewis, Meriwether, 138
Lexington, Missouri, 125
"Liberal Republicanism," 47–48
Liberia, 65, 70
Libraries, 159, 163
Liggett and Myers, 129
Lincoln, Abraham, 39, 43, 46, 92
Lindell, Peter, 109, 112
Lindheimer, Jacob, 150–51
Linguistic groups: accommodation of, 95; Anglophone, 85; French, 79–80; German, 85–86, 94
Lion of the West, The (Paulding), 193, 197, 202–5
Lisa, Manuel, 107
Literature: boosterism in, 234–36; captivity-narrative genre, 222–23; in *The Mirror*, 236–37; poetry, 223, 228–29; reports during Civil War as, 231–32; in *Reveille*, 228–30; role in shaping St. Louis, 237–38; travel narratives and journals as, 2, 218–21, 225–26. *See also* Newspapers
Loan, Benjamin, 47
"Loan Office," Missouri, 106
Localism, 50
Long, Edward H., 180, 183, 185–86
Long Jakes (Deas), 212, *213*
Louisiana Purchase, 25
Louisiana Purchase Exposition. *See* World's Fair (1904)
Louisiana Territory, 53–55
L'Ouverture, Toussaint, 53
L'Ouverture School, 186

Lovejoy, Elijah, 60, 225
Lower Baptist Church, 72
Lucas, Charles, 115n36
Lucas, James H., 111, 133–34; bank of, 123–24; and railroads, 116, 119, 128, 131
Lucas and Turner bank, 123–24
Luders, Friederick, 151
Ludlow, Noah, 192–95
Lutherans, publications by, 94
Lynchings, 60, 74
Lyon, Nathaniel, 39–40

Majority-minority switches, 79
Mallinckrodt, Emil, 82–83
Mallinckrodt, Julius, 80
Manifest Destiny doctrine, 148
Mann, Horace, 167
Manufacturing, in St. Louis economy, 103, 124–26, 129, 131; effects of Civil War on, 127–28; and need for railroads, 119–20
Marble, Dan, 197, 217
Markets, for St. Louis, 50, 119, 126–29. *See also* Trade
Marquard, Henry, 129
Marquette, Jacques, 218–19
Mary Institute, 32
Masquerade, 209–10
Mastodons, in early museums, 139n6, 140
McCune, John S., 133
McCune and Gaty, 129
McDowell, Joseph Nash, 163
McDowell Medical College, 163
McIntosh, Francis, 60
McKay, A. J., 129
McKee, William, 130
McKenzie, Kenneth, 107
McKinstry, Justus, 127
McNair, Alexander, 28
Meachum, John Berry, 62
Mechanics' Bank, 125
Meier, Adolphus, 129, 132–33
Memphis, Tennessee, 115
Merchants. *See* Business
Merchants' Bank, 125
Merchant's Exchange, 112, 127
Metals processing, in St. Louis economy, 126
Mexican War, 35–36
Mexico, trade with, 108
Midwest, St. Louis as distribution center for, 125
Militia, 44, 89; in Camp Jackson incident, 40, 230–31
Miller's Exchange, 112

Mineralogy, 144
Mining, 51, 53, 144
Mirror, The, 236–37
Mississippi River, 84, 103, 127; bridge across, 49, 131; in literature, 223–25; shipping on, 106–7, 113, 117
Mississippi Valley Agricultural and Mechanical Fair, 121–22, *122*
Missouri, 33, 43, 45, 56, 109; black codes of, 55, 57; free black population of, 55, 63; funding railroad construction, 114–17; relations with St. Louis, 20, 26, 32–35; slaves in, 66–67; slave trade of, 58–59; state debt, 117, 120; statehood of, 25, 56; Unionism in, 39, 44–46. *See also* Constitution, Missouri; Legislature, Missouri
"Missouri" (Trollope), 231
Missourian Lays and Other Western Ditties (Umphraville), 223
Missouri Botanical Garden, 5, 32, 166; development of, 157–61; in 1874, *6*; publications of, 161–63; Shaw's hopes for, 233–34
Missouri Compromise, 25, 56
Missouri Democrat, 39
Missouri Historical Society, 2, 32
Missouri River, 103, 106–7, 113, 117, 223
Missouri School for the Blind, 32
Monetary reform, 26
Mordecai, Samuel, 65
Morgan, E. D., 134
Morton, Samuel G., 141, 147
Mount, William Sidney, 193, 210–12
Mullanphy, John, 89, 105–6, 109, 115n36
Mullanphy family, 80
Murphy wagons, 108
Murray, J. H., 73, 74
Murtfeldt, Mary, 153n54
Museums, 136; buying plant collections, 151–52; evolution of, 138–40; Shaw's, *160*
Music, 75–77, 201, 210
Myers, Drummond, and Catlin, 129
Mysteries of St. Louis, The (Boernstein), 42, 94

Narrative of William Wells Brown, a Fugitive Slave, Written by Himself, 225–27
Nash, Pelagie, 65
National Banking Act, 125
National Institute, 148
Native Americans, 51; accounts of captivity by, 222–23; artifacts in museums, 136, 138; influence on fur trade, 107–8; trading expeditions to, 19–21

Nativism, 43, 79, 211; in St. Louis, 87–89; in Whig party, 42, 89
Negro Jefferson Club, 75
Negro Life in the South (Johnson), 215
Neighborhoods: black, 75, 78; German, 84–85
New England, 29, 132–33
New Orleans, 5; importance of, 19, 84; mythmaking about defense of, 194–95; population of, 130–31; and trade, 20, 106–7, 113, 115
Newspapers: anarchist and socialist, 94; black, 75; Czech-language, 94; Democratic vs. Republican, 87; destroyed by Great Fire of 1849, 227–28; founded by immigrants, 80; French, 80; German, 82; German vs. English, 94; support for state funding for railroad construction, 114
"New St. Louis," 13–14
New St. Louis Theater, 216
"New West," 25
New York, 89, 125, 132–34; in trade competition with St. Louis, 119, 121
Nicholson, David, 133
Nicollet, Joseph N., 146, 149
Noonan, Edward, 91
North Carolina, 34
North Missouri Railroad, 117, 119–20, 131–32
North Side neighborhoods, 84
Nostalgia, 2; in genre art, 214–15; in literature, 228–30; in theater, 215
Nullification crisis, 35

O'Fallon, John, 121, 133–34; and banks, 109, 123–24; and railroads, 116–19, 131; support for science, 145, 156; wealth of, 115n36, 119
O'Fallon Polytechnic Institute, 119
Ohio, 89; Cincinnati, 113–14, 126, 130–31
Ohio and Mississippi Railroad, 114, 119, 123
Ohio River ports, 106–7, 113
Orphanages, 68
"Ouster Ordinance," 47
Overstolz, Henry, 91

Pacific Railroad, 115, 128–29; bonds for, 116–17; construction of, 119–20; Southwest Branch of, 117, 120
Page, Daniel, 112, 123, 132, 134; and railroads, 116, 119, 131
Page and Bacon bank, 119, 123–24
Pageant and Masque of 1914, 14, 209–10
Palmatary, James T., *105*
Panics, financial. *See* Financial panics and depressions

Paramore, J. W., 130
Partridge, George, 132
Party politics, uniqueness of St. Louis's, 86–87
Patriotism, in theater productions, 194
Patronage jobs, 45, 73–74
Paulding, James K., 193, 203–4
Paxton, John A., 3, 104–5
Paxton, Joseph, 5
Peale, Charles Willson, 138–39
Pennsylvania, 89, 125
Philadelphia, Pennsylvania, 125
Philanthropy: of O'Fallon, 119; of Pulitzer, 233–34; of Shaw, 1, 233–34
Physicians, 140–41, 148–49, 153n54
"Planters' House, The" (Grierson), 232
Planters' House Hotel, 225, 232, *233*
Plessy v. *Ferguson*, 73
Poe, Edgar Allan, 228
Poetry, 223, 228–29
Police, Irish in, 92
Political leadership, St. Louis, 19–20, 131–34
Politics, 22, 28; Benton in, 23–24, 36–37; blacks in, 69–71, 73–74, 78; depicted in art, 212, 214; Shaw's lack of involvement in, 5–6; uniqueness of in St. Louis, 86–87
Polk, Trusten, 37
Polytechnic Building, 163–64
Poor, education for, 179, 186
Pope, Charles, 156
Pope Medical College, 163–64
Post-Dispatch, The, 233
Power of Music, The (Mount), 211–12, 214
Pratte, Bernard, 106–8
Presidential campaigns, 43, 48
Presidential election of 1876, 71
Primm, James Neal, 235
Property: of free blacks, 52–53, *54,* 65, 68; rights, 44, 46, 57–58; of women, 115n36. *See also* Real estate
Prosser, Gabriel, 55
Pulitzer, Joseph, 94, 129, 232–33
Putzel, Max, 237–38

Race: in art, 211–12; in plays, 203–4
Race relations: effects of slaveholders' fear of rebellion on, 57–58; between Germans and blacks, 92–93; between Irish and blacks, 93–94; majority-minority switches, 79
Radical movement, 44–47, 71
Radical Republicans, 169, 173
Ragtime music, 75–77
Railroads, 36; connections to St. Louis, 49, 104, 114–17, 128, 131; construction

of, 119–20; corruption in ownership of, 128–29; freight shipped on, 114, 119, 128–30; investment in, 117, 123, 131–32
Rankin, C. C., 75
Real estate, 115*n*36; trade in, 105–6, 111; values of, 105–6, 126. *See also* Property
Reavis, Logan Uriah, 49, 132, 234–36, *235*
Reconstruction period, 47–48, 67–71, 169
Reed, Luman, 211
Reedy, William Marion, 13, 221, 236–38
Relief, for exodusters, 71–72
Religion. *See* Catholics; Churches
Renault, Phillipe, 51
Report on a Journey to the Western States of North America (Duden), 82, 225–26
Republican party, 75, 87; Blair and Brown's move to, 42–43; and civil rights for blacks, 72, 74; German support for, 89, 91–93
Reveille, 227–30
Revels, Hiram K., 62–63
Rice, Thomas D., 201–2, 205–9, *206*, 217
Riley, Charles V., 164
Riots: Camp Jackson incident, 40, 230–32, *231;* of nativists vs. immigrants, 211
Rivers: and St. Louis as transportation hub, 113–14, 117. *See also* Mississippi River; Missouri River; Ohio River ports
Robinson, C. K., 75
Russell, William, 115
Rustic Dance after a Sleigh Ride (Mount), 211

Santa Fe Trail, 108
Schlesinger, Arthur, Sr., 10
Schofield, John McAllister, 66
Schofield Barracks, 66
Schurz, Carl, 47–48, 91, 93, 130
Science, 149, 153*n*54, 157, 163; increasing interest in, 143, 148; support for, 138, 153, 164–65, 166
Scientific societies, 140–43, 154–55
Scott, Dred, 60, *61*, 115*n*36
Secession, 39, 127
Second Great Awakening, 29
Second Missouri Infantry of African Descent, 66
Second Ward, immigrants in, 85
Segregation, 68, 75, 93, 192; in education, 73, 173, 179; lack of uniformity of, 77–78
Senate, U.S., 48; Benton in, 22–23, 25–26, 36–37
Senckenberg Institut, 146–47
Senter, William, 130

Settlers: effects on economy of St. Louis, 103, 106; land sales to, 25
Seward, James, 60
Seymour, Horatio, 48
Shaw, Caroline, 4, 115*n*36
Shaw, Henry, 1, 3, 14, 79–80, 133, *158, 165;* development of garden of, 157–61; finances of, 4–5, 111–12, 149; honoring Engelmann, 165–66; mausoleum of, *15;* on philanthropy, 233–34; social life of, 3–4, 32
Shaw, Sarah, 4
Shaw's Museum, *160*
Shepard School, *178*
Sherman, William Tecumseh, 47, 124, 230–31
Short, Charles W., 151
Silver, 51, 108–9, 112
Silver Dollar Saloon, 75–77
Slaveholders: fear of rebellions, 55, 57; feelings toward free blacks, 65; free blacks as, 62; influence in Democratic party, 43
Slavery, 44, 225; abolished, 67–68; divisiveness over, 126–27, 134, 230–32; in Missouri, 51–52, 56; in Missouri politics, 36–37; Missouri's constitutional amendment against, 46–47, 92; and Missouri statehood, 25, 56; opposition to, 34, 36–39, 42; and responses to *Uncle Tom's Cabin,* 216–17; Shaw's lack of protest of, 5–6; spread of, 35, 52; in St. Louis, 31, 33–34
Slaves, 69, 113; education for, 62–63, 66–67; free blacks sold as, 59–60; labor of, 56, 104; methods of emancipation, 60–62; population of, 53–56, 125; restrictions on, 55–58; runaway, 59–60, 66; in rural Missouri, 34–35; trade in, 58–59, 225–27; treatment of, 52, 57–58
Slavic immigrants, 99
Smith, Jedediah, 107
Smith, John B., 109
Smith, Solomon, 192, 194, 215, 229
Smith Academy, 32
Smithsonian Institution, 148
Social class, 104, 210; depicted in plays, 197–99, 203, 207, 209; and theater, 191–92
Social identities: authenticity of, 201; in genre art, 210–15; in public masquerade, 210
Socialists, 93, 94
Social order: benefits of education in, 168–69, 173–74, 179; disorder of, 20, 26, 32, 33; education as democratizing, 172, 175
Soldan, F. Louis, 180, 184
Sonnenschein, Solomon, 95–96

South, the, 39, 71; demand for slaves by, 58–59; influence on St. Louis political leadership, 132–34; rural Missouri sympathy for, 34, 36, 89; St. Louis as, 127, 134. *See also* Confederacy
Southern Bank, 125
South Side neighborhoods, 84–85
Southwest, the, 131, 148, 152; St. Louis trade with, 108, 118, 120, 128
Southwest System, 129
Spain, 21, 53, 55; colonialism of, 24, 51
Spaunhorst, Henry, 92
Staehle, Johanne, 121
Stanton, Edwin, 44
Star Mill, 112
St. Dominique, 51. *See also* Haiti, black rebellion in
Steamboats, 48–49, 106–7, 113, 227
Stephens, Lon V., 74
Stewart, Robert M., 120
Stewart, William Drummond, 145, 151
St. John Nepomuk, 94
St. Joseph, Missouri, 125
St. Louis, 79; bird's-eye view of, *105;* boosterism of, 19–20, 49–50, 115–16, 234–36; city and county, 117, 131–32; city-county split, 50, 130; civic visions of, 233–36; destiny of, 10, 50, 132; ethnic character of, 40–42, 79, 80–83; founding of, 21, 209–10, *220;* growth of, 2, 19–20, 32, 50, 121; image of, 31, 60, 221–25, 228; importance of, 7–8, 19–21, 34, 143–44, 164; incorporation as city, 14, 103; influences on, 51, 79–80; investing in railroads, 114, 116–17, 131–32; North-South vs. East-West fixations of, 131–32; party politics of, 86–87; past vs. future of, 13–15; prominence of, 75–77, 168–70, 217; riverfront of, *4;* rural Missouri relations with, 26, 32–35; status among cities, 98–99, 130–31; as village, 1, 2. *See also* St. Louis population
St. Louis, Iron Mountain, and Southern Railroad, 129
St. Louis, the Future Great City of the World (Reavis), 132, 234
St. Louis and Iron Mountain Railroad, 114, 117–20, 129–30
St. Louis Association of Natural Sciences, 140–43
St. Louis Brown Stockings, 134–35
St. Louis Cotton Compress Company, 129–30
St. Louis Directory and Register (Paxton), 104–5

St. Louis Enquirer, 25
St. Louis Gaslight Company, 111
St. Louis Hegelian movement, 97, 170
St. Louis High School, 171
St. Louis Labor (newspaper), 94
St. Louis Museum of Fine Arts, 32
St. Louis Palladium, 75
St. Louis population, 19, 33, 121; decline of, 106; free people of color in, 52–53, 59, 125; German immigrants in, 34, 40–41, 83–85, 113, 121; growth of, 103, 113, 125, 153; makeup of, 8, 83–84; overestimation of, 49–50, 105, 130–31, 235; rivalry with Chicago over, 49–50, 130–31, 235; slaves in, 52–56, 125
St. Louis school of philosophy, 170
St. Louis Theater, 192, 216*n*33
St. Louis University, 141
Stoddard, Amos, 55, 221, 237–38
Stowe, Harriet Beecher, 215–16
St. Paul's AME Church, 72
Sts. Peter and Paul Roman Catholic Church, *81*
Stupp, Johann, 121
Sublette, William, 107–8, 109
Synagogues, 94

Tallmadge, James, 56
Tandy, Charlton H., 71–72
Taxes, 105, 114
Taylor, George R., 127–28, 132–33
Tennessee, migrants to St. Louis from, 34, 89, 103–4, 106
Texas, 35
Theater, 42, 217; acceptability of, 191, 215–16; audience role in, 192–93, 197, 215; blackface in, 193–94, 201, 205–9, *206,* 212, 215; influence on art, 211–12; Kentuckian character in, 195, 202–5; social class and, 191–92, 215; of social representation, 205–9
Thomas, James, 65, 231
Tice, John, 172–73
Tobacco, in St. Louis economy, 129
Torrey, John, 150, 152
Tower Grove Park, 5
Trade, 127; in cotton, 129–30; importance to St. Louis, 125, 128, 132; quantities through St. Louis, 112; via railroad, 114, 119–20; via steamboats, 106–7. *See also* Business; Fur trade
Trading expeditions, 19–21
Transactions (Academy of Science journal), 161–63

Transcendentalist movement, 29
Transportation, St. Louis as hub of, 103, 106, 113–17
Trelease, Mrs., 15
Troen, Selwyn, 96–97
Trollope, Anthony, 231
Turner, Harry, 14
Turner, James Milton, 70–73, 74, 75
Turner, John, 72
Turnverein, 97, *98*
Turpin, "Honest" John, 75–77
Turpin, Tom, 77
Twain, Mark, 229
Tweed, William Marcy, 36

Umphraville, Angus, 223
Uncle Tom's Cabin (Stowe), 215–17
Underground Railroad, 59
Union/Unionism, 6, 20; attempts to keep Missouri in, 39; factionalism in, 6, 44–46, 127; of German immigrants, 89, 91; support for, 35–36, 42, 89
Union Merchant's Exchange, 127–28
Unitarianism, 29–31
United States, 22, 165, 191; acquisition of Louisiana Territory by, 21, 53
Universities, 165. *See also* Washington University
Upper Louisiana, 21, 51–53, 221
Upper Midwest economy, St. Louis in, 19–20, 50
Upper South, migration from, 34

Vallé, Jules, 119, 124
Van Buren, Martin, 26
Vashon, George B., 74
Verdict of the People, The (Bingham), 212
Ville, The. *See* Elleardsville
Violence: against blacks, 68, 74; against immigrants, 88. *See also* Riots
Virginia, 133; migration to St. Louis from, 34, 55, 89
Von Humboldt, Alexander, 132, 161
Von Phul, Henry, 80
Vote: for blacks, 48, 69–71, 92–93; for immigrants, 95; property qualifications for, 28

Wahrendorff, Charles, 80
Walsh, Edward, 109, 123–24, 132–33
Ward, John, 194–95
Warner, William, 74
War News from Mexico (Woodville), 212

War of 1812, 21–22
Warrick, Alfred, 60
Washington University, 32, 153, 165–66, 183
Wealth/wealthy, 85, 126, 144; guaranteeing banknotes, 110, 123–24; uses of, 116, 119
Weber, Adna, 9–10
Wells, Erastus, 132–33
Wells, Rolla, 75
West, the, 35, 144, 150; in art and theater, 203, 212, *213*, 217; artifacts and specimens from, 136, 138–39, 155–56; economy of, 28, 130, 191; explorations of, 148–49, 152–53; importance of, 19–20
Western Academy of Natural Sciences: collaborations and connections of, 146–48; decline of, 148–49; finances of, 145; formation of, 143–44
Western expansionism, 19, 26, 35, 43
Western Sanitary Commission, 32, 67–68, 163
"What's the Matter with St. Louis?" (Reedy), 221, 236–37
Wheat/flour, in St. Louis economy, 112
Wheeler, John W., 74, 75
Whig party, 24, 36; Bates in, 29, 43; ethnocultural background of, 86–87; nativism in, 42, 89
Whiskey Ring scandal, 129, 130, 133
White, Albert, 65
Wilde, Oscar, 236
Wilkinson, James, 21–22
Williams, Caroline, 68
Wilson's Creek, battle at, 40
Wislizenus, Adolph, 149, 152
Withnell, John, 133
Wolverton, Forrest E., 223
Women: education of, 153n54; influence on St. Louis economy, 103, 115n36; property of free black, 52–53, *54*, 65; as teachers, 169–70, 177
Wood, Obadiah M., 186
Woodville, Richard Caton, 212
Woodward, Calvin M., 130, 169–70, 182–85, 187
Woodworth, Samuel, 193, 197
World's Fair (1904), 2, 122, 237–38
Wyman, Edward, 156

Yankee character: in art, 211; in theater, 193, 197–201, 209
Yankee Merchants and the Making of the Urban West (Adler), 134
Yeatman, James, 112, 116, 124, 132–33
York, 53